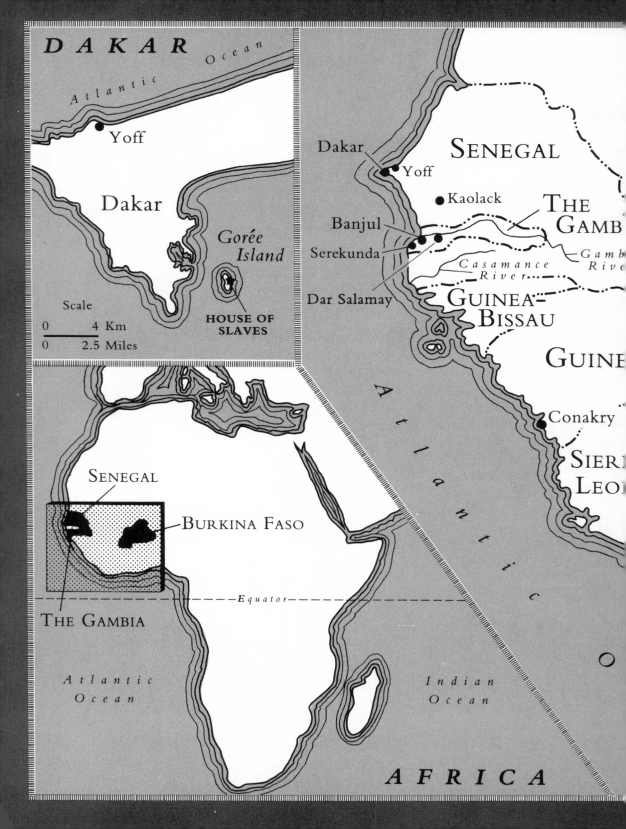

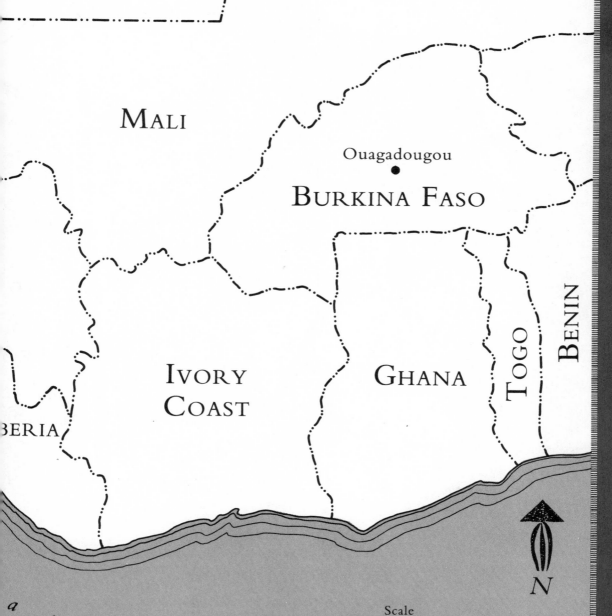

# WARRIOR MARKS

ALICE WALKER *and*
PRATIBHA PARMAR

# WARRIOR
# MARKS

*Female Genital Mutilation and*
*the Sexual Blinding of Women*

*A Harvest Book*
*Harcourt Brace & Company*
*San Diego    New York    London*

Authors' note: To protect the privacy of some of the people who appear in this book, their names have been changed.

Copyright © 1993 by Alice Walker
Copyright © 1993 by Pratibha Parmar
Compilation copyright © 1993 by Alice Walker and Pratibha Parmar
Preface copyright © 1996 by Alice Walker
Preface copyright © 1996 by Pratibha Parmar

Requests for permission to make copies of any part of the work should be mailed to: Permissions Department, Harcourt Brace & Company, 6277 Sea Harbor Drive, Orlando, Florida 32887-6777

Library of Congress Cataloging-in-Publication Data
Walker, Alice, 1944–
Warrior marks: female genital mutilation and the sexual blinding of women/Alice Walker and Pratibha Parmar.—1st Harvest ed.
p.    cm.—(A Harvest book)
Originally published: 1993. With new pref.
Includes bibliographical references.
ISBN 0-15-600214-0
1. Clitoridectomy.   2. Infibulation.   3. Female circumcision.
4. Abused women—Africa.   I. Parmar, Pratibha.   II. Title.
[GN484.W35   1996]
392—dc20   95-25721

The speech on pages 20–26 appeared in the *Radcliffe Quarterly*, September 1992. Copyright © 1992 by Alice Walker. Used by permission of the author.

(Additional permissions acknowledgments appear on page 373, which constitutes a continuation of the copyright page.)

Designed by Lydia D'moch

Map by Richard Knauel

Printed in the United States of America
First Harvest edition 1996
A B C D E

*Warrior Marks*
　　is dedicated
　　to the female child
of Africa
　　in gratitude
for her
　　vulnerable
loveliness
and
　　as a token
　　of my devotion,
which is
　　eternal.

*Alice　Walker*

THE LITTLE GIRL on the previous page has a special significance. I saw her on our first day of shooting in The Gambia. The other children had put on a very lively performance and dance to welcome us to their village, and though they were finished, they milled about, a hundred or so of them, looking us over, laughing and talking among themselves. This child stood apart from all this, quietly watching us with her back protectively supported by a tree. I was immediately struck by her. Her little face held my complete attention. Deborah was busily taking photographs from the window of our car. Finally I asked her to take a photograph of this little girl. Even after she'd taken the photograph and we'd started off, walking, down the road, the child's face haunted me. At that moment Pratibha approached. "Look at that little girl," I said. "I am very drawn to her and haven't managed to take my eyes off her since we arrived." Pratibha took one look and laughed. "But, Alice," she said, *"she's a little you."* At that moment I could see it too. It was exactly as if I were looking in a mirror but seeing my three-year-old self. Only one photograph of me at that age used to exist—I don't know where it is now—and I was suddenly able to remember it. This encounter seemed both uncanny and auspicious, for it was not only the child's body that seemed familiar but, through her dark questioning eyes, her wondering soul.

These are the kinds of questions my father taught me to ask, alas. Adam, he would say, What is the fundamental question one must ask of the world? I would think of and posit many things, but the answer was always the same: *Why is the child crying?*

—*Possessing the Secret of Joy*

# CONTENTS

# Alice's Preface to the Harvest Edition

When we deny our experience, we are always moving from something real to something fabricated. To live by this web of legend will always harm us. The truth may be difficult to open to, but it will never hurt us. What a tremendous relief to have the actual truth openly spoken: "There is suffering in this world."

*—Lovingkindness:*
*The Revolutionary Art of Happiness*
by Sharon Salzberg

IF WE ARE BLESSED, the difficult tasks for which we are born present themselves. It is then that we have the opportunity to put into practice all the knowledge, both spiritual and practical, that has come down to us and that, until the moment at hand, we may have only ritually observed. The five years that I gave myself to work on the issue of female genital mutilation has now passed; it was a period in which I researched and wrote a novel, *Possessing the Secret of Joy,* and was executive producer and co-creator, with director Pratibha Parmar, of the film *Warrior Marks: Female Genital Mutilation and the Sexual Blinding of Women,* an experience this book chronicles. It is not possible, I have found, to put a limit on one's response to suffering, and so, through this book, and in other ways, I know I will be involved in the effort to end the suffering caused by female genital mutilation for the rest of my life.

When, as a college student helping to construct a school building in rural Kenya, I first heard whisperings about female genital mutilation, or "female circumcision" as it was inaccurately called, nothing in my experience had prepared me to comprehend what was meant. I am indebted to African women novelists, whose books I voraciously read after I returned to college, for invariably describing a scene that involved a young girl's initiation into womanhood. Implicit in this description was the message that what was done to the girl made her marriageable, "clean." In fact, it was said that the girl, who was about to encounter the life-threatening shock of having part of her body cut away without warning or anesthesia, was about to be given "a bath." Years later, while I was working as a contributing editor at *Ms.* magazine, the subject came before me again as feminists relentlessly scrutinized the historical and global oppression of women. Gloria Steinem and Robin Morgan wrote a searing essay about it, "The International Crime of Genital Mutilation," and African and Middle Eastern women wrote to the magazine from time to time to express their personal experiences with the practice and outrage at its continuation.

There is some information which, having been received, is impossible to forget or to pretend that it was meant, perhaps, for someone else. And so, over the years, this subject trailed me. It surfaced in my dreams, in essays, in lectures, and in my novel *The Color Purple*. A minor character in the film based on that book, a young woman from Kenya, reminded me, simply by her presence, that I had unfinished work to do. Something else reminded me, which I realize now I usually forget in explaining how I came to work on the issue of female genital mutilation:

One day, on the "African" set of *The Color Purple* (in the dusty California hills), an African actor, formerly a physician in his own country, handed me Joy Adamson's book *The Peoples of Kenya.* The page the book opened to, and the line my eyes immediately rested upon, described the ritual in blood-chilling detail. I later learned that Joy Adamson had been mysteriously murdered after writing her book, and I believe it was this information, coupled with the knowledge that everything about female genital mutilation has for millennia been taboo, that caused me to drop this episode from memory until recently, when I was rereading my journals from that period in order to create a book about making *The Color Purple* into a film.

Though it is doubtful that I will write more about female genital mutilation or speak about it very much—having done so in scores of cities in the U.S. and abroad and before enormous, generally responsive and thoughtful audiences—I will always, because of *Warrior Marks,* be involved in the eradication campaign, which has been carried on valiantly for decades by others. I found myself quite drained, depleted, after five years of immersion in the subject and have had to take a rest from it; I can only imagine the depth of weariness felt by those whose every waking moment, over decades, has been concerned with how to raise consciousness around the issue, how to influence hearts and minds.

*Warrior Marks* the film has been shown in hundreds of cities across the U.S. as well as on hundreds of college campuses. It has also been shown to women's groups in Africa and to immigrant women and men in many parts of Europe. It has been shown at countless feminist and women's film festivals. It is often the first exposure to the subject of female genital mutilation that many people have. For that reason, it is an invaluable tool in the construction of an informed frame of mind. Exposure to the film has helped create a climate in which women from mutilating cultures can step forward and, in some instances, like that of Lydia Oluloro and her daughters Shade and Lara, effectively argue that the threat of mutilation in their native country—in this case, Nigeria—means that they should be accepted in the U.S. as legitimate refugees. The film's presence, as well as the fact that by now most Americans have some awareness of what female genital mutilation entails, has also supported the legislation introduced into Congress by Rep. Pat Schroeder to make genital mutilation a crime in North America. Both Amnesty International and the United Nations have declared female genital mutilation an abuse of human rights.

What is magical about difficult tasks is the diversity of angels one finds assembled to help one, even if most of them are people you never meet, and even if they have no idea they are helping you. The list in this instance is long, but prominent upon it are: Fran Hoskens of *Women's International News;* Abe Rosenthal of the *New York Times;* Hanny Lightfoot-Klein, an intrepid traveler and questioner and wonderful writer; and Soraya Mire, whose moving film "Fire Eyes" brings a Somali woman's story of mutilation unforgettably to the screen.

In fifty years female genital mutilation will be an unpleasant memory in mutilating cultures, as foot-binding is today in China. By then, those of us to whom it has been of pressing concern will no longer be here to congratulate those who are alive to witness its defeat. I know the custom will end, not, unfortunately, because there will suddenly blossom a new respect for the physical and spiritual sovereignty of women and children, but because even those men in high levels of government will learn, possibly because of the spread of AIDS, which is fa-

cilitated by female genital mutilation, that violence done to anyone, but especially violence done to a helpless child, cannot be
permanently hidden, but will eventually permeate, poison, and
even kill the entire society.

This is not a subject, dear reader, to be approached by the
faint of heart; not, that is, until you have consciously strengthened yourself. We live in extremely challenging times, and care
must be taken to learn, in ways that are not self-annihilating,
many of the cruel things we are destined to be told. This is the
time, nonetheless, for which we were born. So sit down, have a
cup of tea, take deep breaths before extending your hand.

Our comfort, as always, is that we are here, in this time,
together. And that we can, if we try, open to each other. To do
this already leads us out of our misery and fear and loneliness,
our sense of not knowing what to do. We soon learn we do not
have to go anywhere, really, to start trying to lessen suffering;
we can begin where we are.

*September 1995*

# Pratibha's Preface to the Harvest Edition

||||||||||||||||||||||||||||||||||||||||||||||||||||||||||||||||||||||||||||||||||||||||||||||||||||||||||||||||||||||||||||||||||||||||||||||||||

IN JULY 1995 AMIRA KAMIL, a 14-year-old girl from the Egyptian village of Kufr Tawil, north of Cairo, screamed in agony when the village doctor used scissors to slice off her clitoris and labia minora. She continued to scream from the time she left the clinic until she died a few days later in her mother's arms. The Egyptian Organisation of Human Rights went to court in August 1995 in an attempt to ban the practice, which affects 3,000 young women and girls every day in Egypt.

When Alice Walker and I completed our film *Warrior Marks* in June 1993, we were keenly aware that the film could only be one small contribution to the international campaign to eradicate female genital mutilation. It is now over two years since both the film and this accompanying book were launched both in Britain and in the United States.

The first public screening of *Warrior Marks* was on Channel 4 Television in Britain in October 1993, when over a million viewers watched the film. Subsequently, Alice Walker and I presented the film in Paris and at the Birmingham International Film Festival in England.

In November 1995 Alice Walker and I embarked on a seven-city tour of the U.S. to screen the film and have discussions with audiences. Our U.S. journey started in New York, and over a period of two weeks we traveled to Washington, D.C., Atlanta, Chicago, Seattle, Los Angeles, and San Francisco. This tour was jointly organized by the U.S. distributor of the film, Women Make Movies, and Leigh Haber from Harcourt Brace, the publisher of this book. Primarily conceived of as a "Benefit Tour," the screenings, for the most part, were sold out, and the money that was raised benefited several women's organizations, including FORWARD International.

Some of the other organizations that benefited from this fund-raising effort were the Women's Foundation and the Bay Area Black Women's Health Project (San Francisco), and the Black Women's Health Project and the Black Cinemateque (Atlanta). The Los Angeles Women's Foundation earmarked the money for grants to African-American women's health organizations, including groups working on domestic violence, drug and alcohol abuse, and the homeless. The Women's Funding Alliance in Seattle distributed funds to a coalition of organizations that included the Abused Deaf Women's Advocacy Services, Center for the Prevention of Sexual and Domestic Violence, Domestic Abuse Women's Network, Harborview Sexual Assault Center, King County Sexual Assault Resource Center, Lesbian Resource Center, Northwest Women's Law Center, Seattle Rape Relief, Washington State National Abortion Rights Action League Foundation, and the Welfare Rights Organizing Coalition.

Since then, the film has continued to be screened at educational and health institutions, theaters, women's events, and film festivals around the world, including those in Senegal, Gambia, Italy, Spain, Norway, South Africa, and Germany.

*Warrior Marks* has evoked a range of emotions, thoughts, reviews, newspaper and journal articles, op-eds in mainstream newspapers, and even articles and letters on the Internet. The film and the book have acted as catalysts for discussions on the subject of female genital mutilation from journals and newsletters with limited circulation to newspapers and magazines with international distribution. This incredible breadth and depth of coverage can only be positive. They have helped to raise awareness not only of the practice of female genital mutilation but also of other practices of violence against women worldwide. The widespread attention and publicity that the film and the book received have also generated a groundswell of activism against the practice of female genital mutilation.

Surprisingly, there was much less controversy than I had expected, given that this subject has always raised passionate responses. It has been clear that not all women agree on the best strategy for campaigning against female genital mutilation, as shown by the variety of responses to the film and this book. For instance, our North American distributor, Women Make Movies, had one African-American woman filmmaker demand that they censor the film and take it out of circulation, while on the other hand, we received letters from and met with many women of all races and cultures who expressed relief that they had found a forum to talk about their experiences of mutilation. Most important, the film's high public profile has encouraged many women who have remained silent about the pain of their own mutilations to begin to share their anger and pain and to end their isolation. Some of these women have become strong advocates for the campaigns to abolish female genital mutilation.

In the last two years when I have traveled with the film to screenings in both the U.S. and Europe, I have sometimes had African, Asian, and Middle Eastern sisters voice their relief that the taboo is finally being dismantled and that the public consciousness around this issue has eased their loneliness. What has become clear is that there is a huge lack of counseling support structures for many of the women who have begun to confront the pain of their childhood mutilations. There are very few places for them to go for refuge or for the ongoing support that they need.

There is no doubt about the agreement amongst many grassroots women activists that the international debates the film and the book have stimulated and the attention that has been brought to their campaigns have been hugely beneficial. These are the women who travel from village to village holding health workshops, talking to circumcisors, and pressuring their governments to enforce and/or introduce legislation to eradicate female genital mutilation. These are also the women who have for so many years been working silently and with very few resources to raise international awareness of this issue.

I believe that Alice Walker's novel *Possessing the Secret of Joy* and the film and the book *Warrior Marks* have been instrumental in opening up the debate on female genital mutilation among opinion makers and in ensuring that international agencies put this issue on their agendas.

In the last two years we have seen international agencies, particularly those based in the U.S., commit money and resources to work against female genital mutilation by creating consultancy and advisory posts for women. It remains to be seen how these resources will filter down to the grassroots activists who are at the forefront of efforts to abolish this harmful practice.

After the Channel 4 broadcast, students at Sheffield University in England put together a petition to the British section of Amnesty International demanding that the organization accept female genital mutilation as an act of torture. They used both

*Warrior Marks* and *Possessing the Secret of Joy* as supporting evidence. Consequently, Amnesty International accepted the students' demand.

A recent development took place at the United Nations Fourth World Conference on Women in Beijing, China, where the final document produced by women representatives from over 136 countries around the world, "Platform For Action," specified "that women should have access to protection against all forms of violence, including rape, genital mutilation, domestic battering and sexual harassment." This is a major achievement given that when women tried to raise the issue of genital mutilation at the first United Nations women's conference in 1980 in Nairobi there had been an outcry and a reluctance to discuss this issue.

Change is slow particularly in countries where resources are low and governments are reluctant to adopt new laws or practices to expand women's rights, but every effort to eradicate violence against women, and genital mutilation of girls and women in particular, helps the global and local campaigns.

I have no illusion that one small documentary film can answer all the questions or elucidate all the historical, political, and economic complexities surrounding the question of female genital mutilation. I hope many more films will be made, capturing the strength and pain of the stories of women warriors and survivors all over the globe.

As we approach the millennium, I continue to commit myself in my work and in my life to speak out against violence against women in all its forms worldwide, violence which sadly shows little sign of abating.

*September 1995*

# PART ONE

# INTRODUCTION

I BELIEVE WE ARE DESTINED to meet the people who will support, guide, and nurture us on our life's journey, each of them appearing at the appropriate time, accompanying us at least part of the way. I think specific human beings, sometimes only in spirit, will present themselves in such a way that their presence will shape and reshape our hearts until we are more fully who we are. This particular magic or synchronicity is activated by something both simple and profound: we must adhere to our own peculiar way, that is the only chance we have to meet those spirits who wander along *our* road; we must persist in being true to our most individual soul.

I have observed this process over and over in my life, always with the greatest amazement. Part of the optimism I am able to maintain in the face of the worst realities is a result of knowing that many companion spirits are patiently waiting to join me in any endeavor or along any path I take in the world—after all, this has simply been my experience so far—and that all I must do is begin.

There is a feeling of security in this and a degree of serenity that astonishes even me.

It was in this mode—my elders would call it the mode of faith—that I approached Pratibha Parmar about the possibility of the two of us creating a film about female genital mutilation, a subject so heartrending it had already given me a new appreciation of the word *grief.*

By the time I asked her, I had already seen her films *Sari Red, Khush,* and *A Place of Rage* and knew she had her own way, her own vision, her own light. Yet it was clear she had respect for the input of others (some of her "actors" in *Sari Red* were friends and family members) and could be flexible in the creative matrix of the collective effort. Besides, we became friends very quickly and were almost immediately able to discuss bluntly any and everything, without taking the trouble of beating around anything at all. This ability to cut through to the matter at hand and to "fly" together in thought and conversation is one sign, I think, that a companion spirit has been met.

After working on *Warrior Marks* with Pratibha for nearly a year we are, I think, especially grateful that close association has not curtailed our mutual respect. Though each of us suffered in her own way while making this film the experience has had the result of deepening our friendship. We dream of collaborating again in the future.

When I saw the completed *Warrior Marks,* I recognized it as a symbol of our mutual daring and trust. It is a powerful and magnificent film, thanks to Pratibha's brilliance as director, constructed from our grief and anger and pain. But also from our belief in each other, our love of life, our gratitude that we are women of color able to offer our sisters a worthy gift after so many centuries of tawdriness, and our awareness of those other "companion spirits" we know are out there.

Perhaps you, reading this now, are one. If so, welcome to this journey. Hazardous. But guaranteed to work the heart into a bolder shape.

# ALICE'S
# JOURNEY

Dear Pratibha,

I've asked Joan* to send you a copy of my new novel, *Possessing the Secret of Joy*. After you've read it you'll understand my present preoccupation. What I'd like to discuss with you is the possibility of doing a documentary (not on the novel, which might or might not become, somehow, a feature film, requiring tons of money) about genital mutilation. I would produce, you could direct. I'd also narrate. We could travel to Africa, etc. etc. I could use some of the money I'll get for the book, and we could hope to air the film across the continent of Africa, beginning in Ouagadougou. I've barely begun to think about this, but I wanted to at least share with you something of what I am thinking. It could be a simulation, similar to the one we watched at the Roxie on footbinding, but also important, I think, would be to have interviews with women, men, and children in Africa. Perhaps in Senegal and Kenya. And several medical opinions. How much money do you think we'd need for this? If it's an hour long?

Let me hear your thoughts on this, please.

My warmest hug to you and Shaheen.

Love,
*Alice*

---

*Joan Miura, administrative assistant to Alice Walker.

*December 18, 1991*

Michael Rudell, Attorney
New York, N.Y.

Dear Michael,

    I'm interested in forming a production company whose name will be Our Daughters Have Mothers, Inc. How do I do this? What is required legally, I mean. And can you do it? This would be primarily for making a documentary about genital mutilation. Not based on my novel.

    Thank you for all your work.

Happy holidays.
*Alice Walker*

---

*To:* Pratibha Parmar
*From:* Alice Walker
*Regarding a film on genital mutilation*
*February 24, 1992*

Dear Pratibha,

    I am off to Australia, New Zealand, and Bali today. I look forward to encountering a new/very old world. I hope you are well. It was lovely being with you, too. Here is the statement you asked for. I hope it is useful. Do change it if you need to.

*I wish to co-produce and narrate a one-hour documentary film about all aspects of genital mutilation. I wish to be filmed on site in countries where the ritual of "female circumcision" is practiced. I wish to visit elders, matriarchs, and young and older married couples and interview them about their experiences—sexual and psychological. I wish to visit hospitals where these "surgeries" are done in Third World countries as well as in Europe, England, and North America. I would like to assist in the filming of a clitoridectomy or help arrange a simulation. I would like to show the resulting appearance of a woman who has been pharaonically circumcised and infibulated.*

*My hope is that following publication in the U.S. and England of my novel* Possessing the Secret of Joy *(which explores the life of a genitally mutilated woman), there will be a window of opportunity to be candid and connected (the book will serve as the tool of connection) regarding genital mutilation, which endangers women and children wherever they live and impacts negatively on people's lives and health around the world. The likelihood that genital mutilation hastens the spread of AIDS will also be explored in this film.*

*Without necessarily condemning anyone, this film will present the facts about the horrors and hazards of a practice that is thousands of years old, with the intent to encourage people to reevaluate their "traditions" in light of the fragile health of their societies and the planet.*

*I believe that education, and a committed openness about discussing the realities, will make the most difference where stopping this particular violence is concerned. I look forward to working with director Pratibha Parmar, whose expertise and background will be invaluable assets to this endeavor. Together we will produce a film that serves the world.*

That's the statement, hastily written. I am tempted to add something I've been thinking about, and perhaps you will want to add it. That in the "enlightened" West, it is as if genital mutilation has been spread over the entire body, as women (primarily) rush to change their breasts, their noses, their weight and shape—i.e., by removal of ribs and fat, and by such things as deliberate starvation. I would want this in the film somehow, because otherwise there will be a tendency for Westerners to assume that genital mutilation is more foolish and "barbaric" than the stuff they do.

Anyway, I hope to speak to you when I come to England or when you're back here.

Lots of love,
*Alice Walker*

||||||||||||||||||||||||||||||||||||||||||||||||||||||||||||||||||||||||||||||||||||||||||||||||||||||||||||||||||||||||||||||||||||||||||||||||||||||||||||||||

*September 7, 1992*

Dear Pratibha,

I am so delighted to hear your good news regarding our film! Rebecca* and I are ill with a virus, but your fax lifted our spirits. We were both much too active this summer and have vowed to become more sensible. I can't respond to everything, but I do want to respond to as much as possible.

1. Great about Channel 4's 80,000 pounds! My 150,000 is in hand. No, I don't at the moment have a conduit for it. If you can work something out there, fine. Rebecca thinks the 60,000 should be fairly easy to raise. Otherwise I could put it up, perhaps further along in the process.

2. Yes. Kenya and Sudan sound just horrible, and my feeling is there's no point in endangering ourselves. Let's think in terms of

---

*Rebecca Walker, Alice Walker's daughter.

West Africa only and use footage from Egypt and Sudan that appears in other films. If in fact we even need this. I think Senegal, Mali, and perhaps Ghana or Nigeria or Ivory Coast would be more than adequate. Cheaper, too. There's also Burkina Faso, fascinating because Thomas Sankara, the revolutionary young president, was ardently opposed to genital mutilation and made a wonderful speech denouncing it. It is said he was assassinated by his closest friend. It is done as well in Sierra Leone. No lack of places to choose from, alas. The woman we need to help us with contacts is Efua Dorkenoo. She resides in London and is head of FORWARD International [Foundation for Women's Health, Research and Development], an organization that aims to eliminate destructive traditional practices. I am hoping she will suggest places and people to visit in West Africa. She is Ghanaian, and I'm sure genital mutilation is practiced there, in Ghana. Would you like to arrange a meeting with her before I come to London? I'm scheduled to meet her on October 14. If yes, I suggest you first introduce yourself to Sarah Wherry at Jonathan Cape. She's my publicist and a very lovely and helpful woman.

3. I have also been in touch with Aminata Diop and her lawyer, Linda Weil-Curiel. I asked if we might interview them in Paris, on our way to West Africa. They are agreeable, in principle. It would be valuable, especially if one of the countries we visit is Mali, from which Aminata fled. That reminds me: I've always wanted to go to Cameroon. Let's consider it, as well. But no more than three countries in all. Two, if we can use footage from other films.

4. If you will be kind enough to draft the letter of intent, I will sign it and send it on to Channel 4.

5. Yes. The October visit. I agree these events should be filmed. I also hope we can do the filming you suggested here in California. I'm formulating a kind of script to be feathered through the film, and part of that would be in Mendocino.

6. Rebecca regrets she will be unable after her strenuous summer to assist us in research. She hopes to do other things. Perhaps fund-raising, though at the moment it makes her head ache to think about it.

7. I'll ask Joan to send you some of everything we have in the way of research. I'm also faxing a kind of bare-bones chart. Efua Dorkenoo's organization puts out great materials, and I'm sure she'd give you a couple of copies of her magazine.

8. The title that sticks with me is "Warrior Marks: Genital Mutilation and the Maiming of Women." In connection with this: I recently saw a British-made documentary on America's obsession with youth, called *Never Say Die*. There was a segment of a woman having a face-lift. Horrible. I'd love to have this in our film. The title will make sense when you see my little script, which will tie genital mutilation to other psychic and physical maimings women endure. It is also a way to make these mutilations visible in a different way, instead of only the pitying/horrified ones.

I'm planning to be here after I return from England and Holland, the last week of October, all of November, and then I'll be in Mexico all of December.

I trust you are well. I had a chat with June today. She sounds strong, but of course her loss is ongoing.

Our party was fabulous. We had a salsa band, great food,

great art (Huichol), and beautiful dancing people who stayed, many of them, through the next day.

See you soon.
*Alice*

P.S. What are the prospects for "Warrior Marks" being shown in the U.S., on PBS? Do you have contacts? Information? I know nobody will want to put up money here if the film is shown only in England.

---

*November 30, 1992*

Dear Pratibha,

I am sending you the little script that I hope will be part of the film. I don't know just how you'll do it, but I think it can be worked in throughout the discussions about genital mutilation, so that I am a part of the subject and not just an observer. I've done this in a deliberate effort to stand with the mutilated women, not beyond them. I know how painful exposure is; it is something I've had to face every day of my life, beginning with my own first look in the mirror in the morning!

This might well change the way we envision our film. See what you think.

I regret very much the ugliness of the format. I can't get anything more graceful out of this computer.

I hope to see you very soon.

Love,
*Alice*

P.S. Pratibha, another important image I'd like to work in is that of the mutilation of hair, which, because of the reasons given for doing it, approximates genital mutilation. Just as mutilating cultures believe firmly in the ugliness of the natural vulva, African Americans overwhelmingly believe in the ugliness of their naturally textured hair. I'd like to see a small girl having her hair straightened by her mother. Holding her ears, grimacing in pain from the heat of the straightening comb. There is also the proliferating use of fake hair, "extensions," which is reaching crisis proportions. Soon black women won't have any idea what their own hair is like. As anyone who has natural "dreadlocked" hair can tell you, many black people are already, because of hair straightening, completely shocked to realize, on touching someone's natural locks, that this springy, long, easily cared for cascade of hair is *their* hair.

There is an economic similarity, as well. Throughout Africa, one can see mothers who barely have enough to eat spending their last cent on synthetic hair and bleaching creams. (They will also spend all their money to have a child mutilated and thus made "attractive" and marriageable.) Since textured hair and black skin are natural to Africans, it is no challenge to see that they wage a futile battle to change themselves, at great pain and expense, into something they're not. In Africa, Michael Jackson is loved as much for his hair straightening and light skin as he is for his music. But all African Americans who change their features and hair send a powerful, near-irresistible message to our sisters and brothers back home, who unfortunately imitate some of our least self-loving traits.

I also enclose a copy of the statement I gave last spring at the Radcliffe (now Bunting) Institute in Cambridge, "The Light That Shines On Me." It expresses many of my feelings about how I view my work on female genital mutilation.

# Like the Pupil of an Eye

|||||||||||||||||||||||||||||||||||||||||||||||||||||||||||||||||||||||||||||||||||||||||||||||||||||||||||||||||||||||||||||||||||||||||||||||||

*Genital Mutilation
and the Sexual Blinding of Women*

**pupil** *n:* pupil of the eye [fr. L *pupilla,* fr. dim.
of *pupa* girl, doll, puppet; fr. the tiny image
of oneself seen reflected in another's eye]

> *Webster's Third International Dictionary*

E x . D a y : Long shot of Alice and her dog, Mbele. They are
approaching the deck of her house in Mendocino.

Now they are on the deck. Alice sits in the swing. Mbele sits
at her feet. Alice is talking to the dog.

C a m e r a : Camera closes in on Mbele's face. It notes: One
eye is brown, "normal," the other gray.

*(Voice-over: Different locations, etc.)*

A l i c e : At Christmas when I was seven years old, my brothers
were brought air rifles by Santa Claus. These guns shot copper
pellets, which were used to kill birds. Because I was a girl, I did
not receive a gun.

A l i c e : The brother who'd always attacked me with threats
and fists before, who was two years older and quite a bit larger,
now used his gun to shoot at me. One day he shot me in the

eye, destroying the pupil. Within minutes I was blind in that eye.

A L I C E : For a long time my eye was painful. For many years it was covered by scar tissue. I have endured many surgeries. I have walked into many objects I could not see.

A L I C E : After caring for me the week it happened, my parents ignored the injury. They referred to what had happened as "an accident." *Alice's* accident. When I failed to adjust to a new school environment because of the resulting handicap and hostile curiosity of my classmates, I was sent away to live with my grandparents.

A L I C E : For a long time I felt completely devalued. Unseen. Worthless. Because I had been blamed for my own injury, yet could not accept that it was my fault, the thought of suicide dominated my life.

A L I C E : The conflict for me was: Although he was only ten, I had seen my brother lowering his gun after shooting me and knew the injury had been intentional. Perhaps he had not planned to shoot me in the eye, but that he was aiming at me was unmistakable. I was standing on top of our makeshift garage. Nothing was up there but me.

A L I C E : One day, after the birth of my own daughter, I confronted my mother. My father had died, never speaking to me about what had happened. After my injury, in fact, he completely withdrew. His own mother had been shot to death when he was eleven, by a man who claimed to love her; maybe the sight of my injury pained him, as it struck that old bruise of loss and fear. Maybe not. In any event, this is something I will never know.

ALICE: My mother asked me to forgive her. She and my father had of course purchased the gun that shot me. It was she, in particular, who had been in love with "shoot-'em-up" Western cowboy movies. She hadn't considered the consequences of buying my brothers guns.

ALICE: What I had, I realized only as a consciously feminist adult, was a patriarchal wound.

ALICE: I remembered how it had been, growing up in the fifties in the segregated South. How hard black people worked and how little white employers paid them. Their only entertainment at the end of an exhausting six-day week was the "picture show." There they consumed racist and sexist propaganda, via the "shoot-'em-ups" my mother loved, which taught them to despise Indians and Africans as a matter of course. To distrust Asians. To protect and respect only white women; to admire and fear only white men; and to become unable to actually see themselves—for the duration of the film, at least—at all. Theirs was in fact a psychic mutilation.

ALICE: The fact that I learned to rebalance, to continue, to go on with my life, without the support of my parents' protection and thoughtfulness, means I have by now turned my wound into a warrior mark—for I have had to live with it and to transform myself, from someone nearly devastated by childhood suffering, into someone who loves life and knows pleasure and joy in spite of it. It is true I am marked forever, like the woman who is robbed of her clitoris, but it is not, as it once was, the mark of a victim. What the woman warrior learns if she is injured as a child, before she can even comprehend that there is a war going on against her, is that you can fight back, even after you are injured. Your wound itself can be your guide.

A L I C E : It was my visual mutilation that helped me "see" the subject of genital mutilation.

A L I C E : Today the maiming and mutilation of women is common. Not just in the Middle East. Not just in Asia. Not just in Africa. A few months ago, my niece, Linda, after being robbed at gunpoint, was told to run. As she did so, she was shot in both legs. She lost one of them. *(Scene of Linda on crutches.)* Certain members of my family continue to refer to this assault as "Linda's accident."

A L I C E : I was not surprised to learn, while doing research for my book *Possessing the Secret of Joy,* that women are blamed for their own sexual mutilation. Their genitalia are unclean, it is said. Monstrous. The activity of the unmutilated female vulva frightens men and destroys crops. When erect, the clitoris challenges male authority. It must be destroyed.

A L I C E : I was eight when I was injured. This is the age at which many "circumcisions" are done. When I see how the little girls—how small they are!—drag their feet after being wounded, I am reminded of myself. How had I learned to walk again, without constantly walking into something? To see again, using half my vision? Instead of being helped to make this transition,

I was banished, set aside from the family, as is true of genitally mutilated little girls. For they must sit for a period alone, their legs bound, as their wound heals. It is taboo to speak of what has been done to them.

A L I C E :  No one would think it normal to deliberately destroy the pupil of the eye. Without its pupil, the eye can never see itself, or the person possessing it, reflected in the eye of another. It is the same with the vulva. Without the clitoris and other sexual organs, a woman can never see herself reflected in the healthy, intact body of another. Her sexual vision is impaired, and only the most devoted lover will be sexually "seen." And even then, never completely.

A L I C E :  It is for this loss, among others, that we must, women and men, mourn. For who among us does not wish to be seen completely? And loved in our entirety?

A L I C E :  Those of us who are maimed can tell you it is possible to go on. To flourish. To grow. To love and be loved, which is the most important thing. To feel pleasure and to know joy. We can also tell you that mutilation of any part of the body is unnecessary and causes suffering almost beyond imagining. We can tell you that the body you are born into is sacred and whole, like the earth that produced it, and there is nothing that needs to be subtracted from it.

# The Light That Shines On Me

||||||||||||||||||||||||||||||||||||||||||||||||||||||||||||||||||||||||||||||||||||||||||||||||||||||||||||||||||||||||||||

*Upon receiving the Radcliffe Medal on June 5,*
*1992, Alice Walker delivered the following speech:*


TWENTY YEARS AGO, I delivered a speech at Radcliffe called "In Search of Our Mothers' Gardens." It was about the way most of our mothers, and my own mother in particular, persevered in life against the many obstacles that threatened to stifle and destroy their creativity. At a symposium after that address, I attempted to draw the attention of the eminent panel to the perils encountered by young women of color in a racist and sexist society, and to the fact that they were killing themselves, committing suicide, at an alarming rate.

I remember bursting into tears of frustration and anger because one of the panelists mistook my concern for these young women's lives as an attack upon men of color, who she said must be supported no matter what they did. There was one woman friend (who recently died of cancer) who admonished me for crying. I should have had more pride, no matter what I felt. Another friend, who would also, ironically, later have cancer but who would survive it, tenderly placed her arms around me and offered welcome words of understanding and support.

I think of those two women now. One dead—strangled, I sometimes think, by her own denials and unshed tears—the other still vibrantly alive, feisty and rebellious, becoming an ever more rageful and compassionate poet.*

I wish I could tell you that things have truly improved for

---

*The poet June Jordan.

most young women of color on the planet. That life has improved for some of them is of course true. However, many things I have learned about the lives of women and girls since that day of public grief twenty years ago have taken me beyond tears, if not beyond mourning and rage.

I have also come face-to-face with another kind of mother than the nurturing, creative one of my address. For many years now, I have studied, have thought about, the mother who collaborates with the destroyer of daughters. The mother who betrays.

In my recently published book, the novel *Possessing the Secret of Joy,* I explore the life of a daughter so betrayed. A daughter whose culture demands the literal destruction of the most crucial external sign of her womanhood: her vulva itself.

This is a brief excerpt from the novel, an exchange between Tashi, the genitally mutilated woman from Africa, and Raye, her African-American psychiatrist:

*As for the thing that was done to me . . . or for me, I said. And stopped. Because Raye had raised her eyebrows, quizzically.*

*The initiation . . .*

*Still she looked at me in the same questioning way.*

*The female initiation, I said. Into womanhood.*

*Oh? she said. But looked still as if she didn't understand.*

*Circumcision, I whispered.*

*Pardon? she said, in a normal tone of voice that seemed loud in the quiet room.*

*I felt as if I had handed her a small and precious pearl and she had promptly bitten into it and declared it a fake.*

*What exactly is this procedure? she asked, briskly.*

*I was reminded of a quality in African-American women that I did not like at all. A bluntness. A going to the heart of the matter even if it gave everyone concerned a heart attack. Rarely did black women in America exhibit the graceful subtlety of the African woman. Had slavery given them this? Suddenly a story involving Raye popped into my mind: I saw her clearly as she would have been in the nineteenth century, the eighteenth, the seventeenth, the sixteenth, the fifteenth . . . Her hands on her hips, her breasts thrust out. She is very black, as black as I am. "Listen, cracker," she is saying, "did you sell my child or not?" The "cracker" whines, "But listen, Louella, it was my child too!" The minute he turns his back, she picks up a huge boulder, exactly like the one that is in my throat, and . . . But I drag myself back from this scene.*

*Don't you have my file? I asked, annoyed. I was sure The Old Man sent it before he died. On the other hand, this was a question he'd never asked me. I'd said "circumcision" to him and he'd seemed completely satisfied; as if he knew exactly what was implied. Now I wondered: had he understood?*

*I have your file, said Raye, tapping its bulging gray cover with a silver-painted nail and ignoring my attitude. I am ignorant about this practice, though, and would like to learn about it from you. She paused, glanced into the folder. For instance, something I've always wondered is whether the exact same thing is done to every woman. Or is there variation? Your sister . . . Dura's clitoris was excised, but was something else done too, that made it more likely that she would bleed to death?*

*Her tone was now clinical. It relaxed me. I breathed deeply and sought the necessary and familiar distance from myself. I did not get as far away as usual, however.*

*Always different, I would think, I said, exhaling breath, be-*

cause women are all different. Yet always the same, because women's bodies are all the same. But this was not precisely true. In my reading I had discovered there were at least three forms of circumcision. Some cultures demanded excision of only the clitoris, others insisted on a thorough scraping away of the entire genital area. A sigh escaped me as I thought of explaining this.

A slight frown came between Raye's large, clear eyes.

I realize it is hard for you to talk about this, she said. Perhaps we shouldn't push.

But I am already pushing, and the boulder rolls off my tongue, completely crushing the old familiar faraway voice I'd always used to tell this tale, a voice that had hardly seemed connected to me.

It was only after I came to America, I said, that I even knew what was supposed to be down there.

Down there?

Yes. My own body was a mystery to me, as was the female body, beyond the function of the breasts, to almost everyone I knew. From prison Our Leader said we must keep ourselves clean and pure as we had been since time immemorial—by cutting out unclean parts of our bodies. Everyone knew that if a woman was not circumcised her unclean parts would grow so long they'd soon touch her thighs; she'd become masculine and arouse herself. No man could enter her because her own erection would be in his way.

You believed this?

Everyone believed it, even though no one had ever seen it. No one living in our village anyway. And yet the elders, particularly, acted as if everyone had witnessed this evil, and not nearly a long enough time ago.

But you knew this had not happened to you?

*But perhaps it had, I said. Certainly to all my friends who'd been circumcised, my uncircumcised vagina was thought of as a monstrosity. They laughed at me. Jeered at me for having a tail. I think they meant my labia majora. After all, none of them had vaginal lips; none of them had a clitoris; they had no idea what these things looked like; to them I was bound to look odd. There were a few other girls who had not been circumcised. The girls who had been would sometimes actually run from us, as if we were demons. Laughing, though. Always laughing.*

*And yet it is from this time, before circumcision, that you remember pleasure?*

*When I was little I used to stroke myself, which was taboo. And then, when I was older, and before we married, Adam and I used to make love in the fields. Which was also taboo. Doing it in the fields, I mean. And because we practiced cunnilingus.*

*Did you experience orgasm?*

*Always.*

*And yet you willingly gave this up in order to . . . . Raye was frowning in disbelief.*

*I completed the sentence for her: To be accepted as a real woman by the Olinka people; to stop the jeering.*


An estimated ninety to one hundred million women in African, Asian, and Middle Eastern countries have been genitally mutilated, causing unimaginable physical pain and psychological suffering. And though one is struck by the complicity of the mothers, themselves victims, as of the fathers, the brothers, and the lovers, even the complicity of the grandparents, one must finally acknowledge, as Hanny Lightfoot-Klein does in the title of her book about genital mutilation in Africa,* that those who practice it are, generally speaking, kept ignorant of its real dangers—the breakdown of the spirit and the body and the spread of disease—and are themselves prisoners of ritual.

I wrote my novel as a duty to my conscience as an educated African-AmerIndian woman. To write a book such as this, about a woman such as Tashi, about a subject such as genital mutilation, is in fact, as far as I am concerned, the reason for my education. Writing it worked my every nerve, as we say in African-American culture about those areas of struggle that pull from us every ounce of creative energy and pull away from us every last shred of illusion. I know only one thing about the "success" of my effort. I believe with all my heart that there is at least one little baby girl born somewhere on the planet today who will not know the pain of genital mutilation because of my work. And that in this one instance, at least, the pen will prove mightier than the circumciser's knife. Her little beloved face will be the light that shines on me.

And so I thank Radcliffe, an institution that has long thought about the lives of girls and women, for thinking of me when considering to whom to give this award. I don't often accept awards I have to leave home to collect. I personally believe all awards and honors should be brought to one's door, like flowers from a florist. But Radcliffe is different. The Radcliffe Institute once gave me two protected years and a beautiful, quiet room in which to write, many years ago. It gave my daughter, Rebecca, many happy days at the Radcliffe Preschool. And it provided a forum for women of color to confront horrifying realities of

---

*Prisoners of Ritual.*

life in the twentieth century: to cry, to comfort, to struggle forward and continue to grow. Because of this, it has become, like so many other places in the world, a very special place to me. A place where even distressing news can be received and held to the light by courageous scholars who have not lost their rage or their poets' hearts. And the messenger, not stoned, is met with open ears and sent on her way with understanding and a hug.

OCTOBER 15, 1992

*October journeys are best! according to Margaret Walker. And so,*
THURSDAY, 5:30 A.M.    BROWN'S HOTEL, LONDON

Wide awake after a sound sleep following an exhausting but
wonderful day spent with Efua (Dorkenoo, head of FOR-
WARD International, the London-based organization that
fights genital mutilation through education and activism),
Ben (Graham), her husband, and the refugee women's community in Tottenham. Efua met me at the door of the Africa Centre—above which she has one of the tiniest offices
I've ever seen—with flowers, kisses, and innumerable hugs,
all of which I returned with pleasure. She is richly brown,
with large warm expressive eyes and a very dramatic manner. After she had introduced me to several young women
who are still attempting to rebalance after having been mutilated against their will, we were off to visit the Tottenham
collective, basically squatters, all of them refugees from

*The refugee women
of Tottenham.*

French- and Arabic-speaking countries, most mutilated, most with children. The women are learning English. Deborah (Matthews) and I were introduced to them and were able to speak briefly before being carried off to a magnificent lunch, cooked by the women and reflecting the delicious traditional cuisine that most of them would still be cooking at home were it not for war, famine, and the complete mismanagement of life and resources by the men in their areas, compounded by imperialism, colonialism, drought, and other acts of a thoroughly pissed-off Nature.

Much to my delight, Efua had arranged for a Yoruba priestess to come and bless us and our work. She came in, black and glowing, beautiful in her earthiness. It was as if the sea or a tree had danced into the room. And she prayed and she danced, and I prayed and I danced back. Her dignity was absolute, as was her grace, and I knew that if she ever does get her bare feet back into their rightful relationship with the earth, the Universe will hear whatever she has to say. I was certainly happy to hear her. This was a reassuring thought as I realized that one of the women in the very crowded room was slowly but steadily working her way toward my handbag, which I'd flung in my chair when I started to dance. I danced over and grabbed it just before she pounced. It is a relief to be reminded not to romanticize suffering. Not to believe poverty engenders nobility, or that people who have nothing are content with that.

Efua has asked me to be Matron of FORWARD International. Actually, she asked me to be Patron; I told her that because I am female, and a feminist, Patron was not a possibility. Getting it, she laughed and laughed. Now you see why I also hate the way everyone calls women "guys,"

I said. It hides women from themselves. And look what has happened to the word *matron* anyhow; you only think of "matron*ly*," and the image is of an overweight, overdressed busybody in a big floppy hat. (Here I paused to remove my own rather large hat.) But, I said, I hereby reclaim *matron,* and yes, I will be Matron of FORWARD International. Oh, Alice, she said. For so long we've needed a figurehead, and you'll do just fine! Of course, you're not directly from Africa. . . . But I smiled and reminded her that this is definitely not my fault. I've warned her I'm no good at all at fund-raising. We'll see. Efua is just *my sister*—I think we even look a bit alike. Same cheeks and noses. And so perhaps my African self is Ghanaian. I seem to gravitate toward Ghanaians and already have a number of them—Ayi Kwei (Armah) and Akuosua and Abena (Busia) and others—as friends and "relatives." In fact, as if she anticipated this feeling of kinship, Efua gave me, within minutes of our meeting, a traditional Ghanaian dress, which I will wear to the London book launch of *Possessing the Secret of Joy.* I gave her a briefcase from the National Black Women's Health Project. It is made of raffia and says "Sister Care" on the outside. I love these briefcases and carry mine, identical to Efua's, everywhere.

Deborah is a great assistant and fine companion (our friendship is by now nearly ten years old). A crackerjack real estate dealer back home in Berkeley, here she's completely supportive of my work—helpful, alert, funny. Tonight she wore her suit and tie and looked lovely. The doorman addressed her as "sir." This made us giggle as we piled into a bright-red British taxicab and sat in its spacious, clean interior on seats that resembled sofas. I am struck by how frequently cabdrivers here, while they wait, read novels.

The other night we watched the debate between Gore and Quayle on C-SPAN, with Admiral Stockdale thrown in for comic relief. We howled when he started talking about the time he ran his own civilization. Too funny for words, and I *loved* Gore. He is a man of a quality rarely viewed. So refreshing to see. I liked the way he says "fixing." "We're fixing to limit a term." So Southern. (And usually I can live without white men sounding Southern!) I liked his courtesy and his barely suppressed anger. Later there were the usual doughboy-esque commentators who claimed he was "wooden" and other such nonsense, but he was just right as far as I'm concerned. I could tell he wanted to move on to fisticuffs with Quayle in the parking lot after the show.

OCTOBER 19, 1992

Last night at Pratibha and Shaheen's, Aminata Diop and her friend and lawyer, Linda Weil-Curiel, came to dinner. I don't know what I had expected—and actually I do have a picture of both of them on a wall of my study—but they were *more* somehow. More real, more warm, more loving, more hurting, more strong. More fragile. More sincere. *More.* I loved them almost at once. Aminata doesn't speak English. My French has lapsed, and of course no one but Aminata speaks Bambara. She will leave me speechless later when she tells me there are no words in her language with which to discuss female genital mutilation. As a courtesy, we'd asked if she'd like to discuss her situation in her own language in our film, for which we'd supply subtitles. Over dinner, we talked of logistical things. How tomorrow's shoot is being planned, what time we are to meet, who are the members of the crew.

Today we all met by the river, a highly symbolic place in spiritual black Southern tradition. Even if it was the polluted though still gracefully beautiful Thames, with London Bridge in the distance. I was dressed in my long woolen coat; the crew were warmly dressed. As was Linda. Aminata wore a very thin traditional-looking pants suit under an enormous parka that threatened to hide her from the camera. Pratibha asked if she could possibly remove it, just during the actual filming. She complied. Her suit is a lovely blue and looks good against her bronze skin, but I worried

*Alice Walker and*
*Aminata Diop, London.*

||||||||||||

31

she'd catch cold. Acting has to be among the most boring occupations, I think. There's so much stopping and starting. So much sheer hanging about. Aminata and I walked over the same bit of riverside for a couple of hours (this will probably amount to a minute or less on-screen), pretending to converse. Deborah and I wrapped her in her coat the minute, each time, the cameras stopped. I took her hand in mine and never let go of her. I could feel her need of a mother, and I offered myself without reservation.

Interviewing her later, in the upstairs of a restaurant, I realized that what I'd felt was true. Because she'd refused to be mutilated, her mother, like her father, disowned her. Her mother was thrown out of the house. Now—Aminata said, weeping—My mother thinks I am a bad person because people from the Mali community in Paris write and tell her so. And this is only because you speak out against genital mutilation, I ask. She says, Yes, for to speak out against it or to resist and run away as I did brands you a prostitute.

I ask her if she thinks her mother could be persuaded otherwise if Aminata went back (a great risk for her because "circumcisions" are often done by kidnapping and force if the person hasn't been brainwashed enough to submit). She thinks not. What if I went with you? I ask. Imagining myself already there, though having a hard time visualizing her mother, her mother's situation, household, or country. But she shakes her head. No.

There are just some people who you know are good. Aminata is one of them. It grieves me that her own mother could doubt this. And not appreciate what a brave, tenderhearted, and loving daughter she has raised. I want to shout across the miles separating us: *Stand up and be a mother, damn it!* Don't make this child suffer when, after all, *she* is right, not the society that enslaves both of you! Instead, Aminata and I clasp hands; she throws her scarf over her face and head to hide her tears from the camera and I just say fuck it and let mine flow.

OCTOBER 21, 1992.   6:00 A.M.

I woke from a dream about feeding myself. Of a black woman in a pink dress appearing at my door and of me, identical to her, handing her a plump round bun filled with raisins. The work I am doing now on genital mutilation feeds me. So much so that I find myself in a state Deborah simply calls "ready." I see beauty, male and female, and hear wonderful music wherever I go. Labi Siffre's music, for instance, which I would never have heard if I had not done a benefit for Spare Rib in London. After grumbling along with Deborah about being saved for last on the show, and around about midnight of what had been a very long day, I was escorted gently to the stage just as this singer I'd never heard of was finishing. And what was he singing? This amazing song that seemed to be coming from his deepest heart, "Something Inside So Strong." A few nights later we listened to Labi again. I knew his song should be at the ending of our film. Proving once again that to be in the right place at the right time is to be reborn, again and again. And that *being* gets you there.

So much has happened. I've met Efua, Ben, Aminata Diop, Linda Weil-Curiel. Hung out with Pratibha and Shaheen—who is *lovely.* And through it all is Deborah. A friend I enjoy as I enjoyed my friends in grammar school.

Funny. Alive. Brave. Thoughtful. Herself to the max. We have great talks and laying-about times. There's always music.

## OCTOBER 23, 1992

Little sleep last night. Perhaps because of the wine I had. Or that I forgot my vitamins and dolomite. Or my day mostly in the car and wandering about in Vanessa Bell's house and Virginia Woolf's garden—which I thoroughly enjoyed. This was my reward for going on a book tour. Rodmell is as I imagined it: the tiniest village, and their house, Monks House, right on the street, with a large vegetable garden behind the house and lots of places to sit in various parts of the garden. Is it my imagination, or do Aquarian spaces always feel familiar? The elm tree under which her ashes and Leonard's were scattered is no more. There are plaques. Hers says: "Death is the enemy" and that she will go unvanquished and unyielding, etc. "The waves broke on the shore . . ." This seems wrong. If she thought death the enemy, she wouldn't have committed suicide. But this is a quote from one of her novels. Leonard gets to make a direct statement as himself, not a quote from his fiction. The two highest qualities, he says, are justice and mercy—and with the mercy, toleration. This is so Jewish. I stroked both of them (their bronzed heads) and thanked them for their example of friendship, fidelity, and truthfulness. Leonard continues to be one of my favorite men. Virginia, a beloved mentor, muse, sister, a crazed Aquarian.

Flying across the Channel to Holland, I was reading *Conspiracy,* about her and her sister. It is obvious to me that her suicide at fifty-nine was connected to and caused by unspeakable child abuse from the age of thirteen, when her mother died—and perhaps it had gone on for years before. She would understand and cheer the work women are doing against female genital mutilation. She'd be delighted with Efua. She'd admire Deborah's suit and tie! Just as she'd empathize to the point of her own madness with the children who are abused. She's a fine ancestor who, though injured, left no stone within her reach unturned. I felt quite restored by my pilgrimage through her former cabbage patch and on toward the Ouse River (which looks more like a canal), in which she ended her fragile but paradoxically strong and vibrant life. Like her apple trees, heavy with apples, her work still blossoms and bears fruit that feeds the Eves and conscious Adams of the earth.

OCTOBER 25, 1992.    AMSTERDAM

Marina Van der Heijden (Dutch editor and publicist) came to the hotel at 11:00, then a male journalist. Then a photographer. Then Nawal El Saadawi! Such a strong, beautiful woman! We embraced and laughed and felt terrific. She is to speak on the panel tonight, along with me and Astrid Roemer. I love these town-hall-style gatherings, which are usually packed with people who have a lot to say. It is especially important to have such a gathering in the Netherlands, because there's an attempt to legalize mutilation here, as an expression of "culture." But would the liberal Dutch legalize the genital mutilation of their own children? No. And that is the answer. Nawal is one of the first women to write about her own clitoridectomy, at the age of six, in

her riveting, extremely courageous book, *The Hidden Face of Eve*. And because she's also a doctor, she brings both emotion and technical expertise to her discussion of what is being done to women, especially in her country, Egypt. Her organization for women was recently banned by the government and its assets turned over to a bogus "women's organization" run by a man. Under Sadat, Nawal was imprisoned. I have the impression she is now in self-imposed exile. All I can think is how much Egypt needs this woman, but the men there are too backward to see it. Her husband sees it. He reminds me of Leonard Woolf. Every statement I've seen by him shows love and support of Nawal.

Soon Astrid Roemer arrived. Also strong, beautiful, and elegant, a politician and a novelist, originally from Suriname. She whisked me off to lunch and an interview at the American Restaurant, where we were seated by a nice window. She asked lovely and loving questions, by which I do not mean soft ones, and so of course we talked of Love and Earth and God and Goddess and Nature and the nature of happiness and how it is possible to be happy when so much is sad. We decided that what many of our ancestors have said for so long is actually true: You have to learn to find joy in the struggle itself. Otherwise you die on the vine, so to speak.

After lunch I came upstairs and washed my hair. When I left England, the guy who inspected my passport said: Your hair's longer. I said, shrugging: It grew. He said: Well, I've heard of spaghetti growing on trees. . . . I said: Our hair is different from yours. Yours has flat follicles. He looked at me in surprise as I smiled and walked away. In the U.S., custom officials sometimes search my luggage because of my hair. Linda Weil-Curiel said the immigration official was abusive to Aminata when they came over to London from Paris to be in our film. Told her to stop chewing her gum (as if she were a child!) and wouldn't let Linda help her by translating English into French. His intent solely to humiliate her. Petty but crushing stuff to someone fighting, as Aminata is, for her personhood.

## Four months later

### FEBRUARY 4, 1993.  LONDON

And so. Today we are to see Sarah (Wherry), Labi (Siffre), and Efua (Dorkenoo). And at the crack of dawn tomorrow we head into our African journey. Pratibha called last night and warned me about Banjul (The Gambia), which, though it is the capital, has no paved streets. Senegal sounds quite predatory—everyone wanting money before they'll do anything. Horrible.

### FEBRUARY 6, 1993.  *Happy Birthday Bob Marley!*

In bed in Banjul! Arrived yesterday and greatly surprised to find it so cool. Last night we had dinner with Pratibha and the crew at a nice Lebanese restaurant. I told Pratibha about being in love. She was very happy for me. We embraced, giggling like children. It is this love that helps sustain me as I get off into this.

There was a full moon (!) when we arrived. That was the most significant thing, after the unexpected coolness. Big and yellow and bright. Amazingly welcoming. We drove through the sprawling village of Serekunda. Very basic, jammed with pedestrians and cars. The area is heavily Muslim. The women, most of them, mutilated as children. This exposure to so much mutilation has caused a mutilation tape to play at odd times in my head—and I see and almost feel the razor descending and slicing away not only labial lips but facial lips and eyes and noses, as well. These fantasies are extremely upsetting and remind me of a similar period when I had fantasies (waking nightmares) that involved self-blinding. I had to struggle to protect myself, my vision, my sight, by saying over and over: I have a right to say what I see. I have a right to *see* what I see. I need not punish myself or be punished for *seeing*.

The hotel is pleasant. New. Right on the coast. Julius Coles (an old friend and another Aquarian, also head of USAID in Senegal) called the minute, literally, that I stepped into the lobby. He wants to be helpful, as does Ayi Kwei (Armah), my friend who writes so beautifully, who lives in Senegal, some five hours away. The clerk handed me multiple messages from both of them. I have an instant impression of them: Ayi Kwei very dark, all African, Julius very light, African American, "African" by choice. Both very much my brothers in their desire to help.

FEBRUARY 8, 1993. BANJUL, THE GAMBIA

An amazing day yesterday. I woke at seven, dressed, and rushed to join Pratibha, Deborah, and crew for breakfast. After quite a long time we were packed up and started for the village of Dar Salamay. Our driver's name was Malign (believe it or not), tall and skinny, with a warm way and a gentle smile. He used to be a policeman. As we started out, he said he wanted to play his favorite music for us. I sat there expecting something Gambian, but no, he put on "Matters of the Heart," by Tracy Chapman, which made me smile. He loves her music and wants to send her a present. We promised to try to deliver it.

We arrived at a house, after an hour or so. It was surrounded by a bare yard and enclosed by a leaning fence. There was a long veranda and many dark rooms, one with

*Malign, our driver and friend in The Gambia.*

‖‖‖‖‖‖‖

a pile of sand in a corner, all with beds, except for a sitting room on the end near the road. A large group of children surrounded our vehicles almost immediately, and what beautiful children they were! Dark browns and blacks, hair in all stages of kempt and unkempt. Large dark eyes filled with interest, curiosity, wonder. Within minutes they had put on a show for us. Completely unselfconsciously. Spontaneous music made with tin cans and sticks. The little girls doing clapping dances, the boys dancing in costume—masks and various draperies—and brandishing sticks. Very joyous. The crew was enchanted. And started to film.

After a while Bilaela arrived. She is my age, much heavier, wearing sporty eyeglasses that look like space gear, which African women seem to like. After a few moments of introduction and chitchat, I felt quite at home with her. An American television news crew, she explained, had just left last week, and she had helped them with the documentary on female genital mutilation they're making. She says she wished we had arrived first, because she signed an agreement with them that means she can't sign one with us. She said her daughters would be able to help us, perhaps. I told her it didn't matter, not to worry about it. I felt fine just to sit beside her and talk. She and her daughters head the movement against female genital mutilation in The Gambia. She also indicated that our honorarium to her was a bit too modest. I said: Well, we are black women, and our resources are not the same as those of the American television network, which is upper-class white and male (even though a white woman had come out to make the documentary). I refrained from offering more money, having been apprised more than once by Pratibha that we must stay within our budget. Channel 4 in London is putting up half the money for the film, I'm putting up the other half, and I've seen how swiftly my half has been dwindling.

Soon I was interviewing Mary, a large, dark sister dressed in lavender and purple, with a huge length of hair facsimile hanging down her back. Her daughter, "little Mary," was being circumcised. Had already been, actually. Was now in the bush and part of the ceremony—the return of the young women to the village—we'd be filming. Why did she do this? I asked. Feeling grateful that *The Color Purple,* the movie, preceded me to Africa, where it had a large, enthusiastic reception, so that people everywhere now greet me warmly (often wearing something purple) and begin chatting away as if I already understand, even if they are speaking Wolof or Mandinka. So far I understand one Wolof word and use it frequently. That word is *wow,* which means yes.

So: Why, Mary, I asked (and I wanted to ask, Why do you call yourself Mary? but that is another film), did you do this to little Mary? Because it is tradition, she replied. Had it been done to her? Of course. How had this come about? Well, her mother had told her they were going to a place where there were many bananas. (She loved bananas.) When they arrived, she was captured by women she'd never seen before, pinned down by them, circumcised, and kept secluded for two weeks. Was she frightened? Yes. Did she feel her mother had betrayed her? Well, at the time, perhaps, but later she understood it was "tradition."

Who did she think was responsible for this "tradition"? The ancestors; the grandparents. Grandfather or grandmother? Grandmother, because it is the grandmothers who must see that it is done. Did she think little Mary would feel betrayed, angry with her? Well, she intended to take her some sweets, and soon she would forget all about what had happened to her. Had she herself forgotten? Yes. But if she could stop this "tradition," would she? She could not hope to stop it, being only a woman. But if she could, by some miracle, stop it, would she? Yes. Why? Because of the pain. So you *do* remember it. Yes, of course. There was suddenly a look in her eyes as if a deep pit within her had opened; I encountered the blankness of terror. I remembered something Efua Dorkenoo had said: that when genitally mutilated women, *en masse,* understand what has been stolen from them, there will likely be a period when they will need every possible psychological support, and there's none, so far. Where are the hospitals, the psychiatric wards, the psychologists and counselors for these women? Certainly not in Africa. I also wondered about the role of food, particularly "sweets," in the African woman's life. Perhaps the overweight women one sees so frequently are still feeding "sweets" to the little frightened and dishonored child inside in an effort to "reward" her for her loss and to make her forget. I suddenly remembered that my own mother, after I was shot and delirious from fever and pain, cooked an entire chicken just for me.

The men around here are blandly gracious, like all slave-masters, I suppose. We paid our respects to three graying patriarchs representing the *marabout* or *alacar* (not sure of the word they were using for chief; something Arabic) of the village, who is away on business. The men sat on chairs, the women on the ground. As a guest, I got a chair but could barely stay in it. The women had to recite in detail, and several times, everything we'd done since arriving in the village. They seemed bored, weary of their own subservience. I hadn't realized no one can leave the village, even guests, without this ritual of being "released" by the chiefs, who are, of course, always men. This answers the question of why women don't run away. And, too, where would they go?

I missed my beloved so much. Every once in a while I'd mentally leave wherever I was for a kiss.

When we arrived at "the bush," I was in a state of dread. Sure enough, underneath a large tree there was loud singing and dancing. On a long mat sat the circumcised girls, "little Mary," at four years old, the youngest among them. Most of the other girls were eight, nine, ten, or eleven years old. One of them held little Mary on her lap. They sat as still as statues as the grown-ups danced around them. Literally kicking up clouds of dust. There is a barren women's club.

These are the dykes, says Deborah. Their role is to dress as clowns and men in order to amuse everyone else. One of them kept shaking her near-naked butt in the children's faces. I don't know if they *are* dykes. I think they are ridiculed and forced to play this entertainer role so that they, as barren women, can have a place in society at all, since a barren woman is considered of no use whatsoever. One woman had drawn large charcoal-colored spectacles on her face; another had a wig made of twigs. I stood behind the children for a while, then moved across from them so I could see their faces. Their eyes were silent, grave, stunned. What is going on? they were all silently asking. Why is everyone else so happy?

FEBRUARY 9, 1993. *My Birthday*

This work on the film is the present I give myself. I'm happy I was born to do it. My beloved just called to sing Happy Birthday to me, while I am still in bed.

I had to stop writing yesterday because I couldn't bear it. The enormity of what they've lost (had taken, by force, from them) will not be clear to these girls until much later in life. For most, it will never be clear. I'd asked "big Mary" about sexual pleasure. You know, I said, that the removal of sexual organs lessens sexual response and destroys or severely diminishes a woman's enjoyment. Well, she replied, my sex life is perfectly satisfactory, thank you very much! (How would you know, though, I thought.) I said a heartfelt Good for you!, slapped her palm, and let it go. At least this group doesn't infibulate. The girls have been robbed of their full capacity for pleasure. Their bodies have

*Dancers from the barren women's club who dress as clowns and men in order to amuse. Dar Salamay, The Gambia.*

been violated, and by the very elders who should be protecting them.

I've never seen such emphatic, noisy, but overall cold-blooded dancing as that performed by the women and a few of the men of the village. (Later, when I mention seeing men, the crew will tell me I was mistaken; everyone in the ceremony is female, though many look and are dressed exactly like men.) Eventually this part of the ceremony ended. Then there was the sacrifice, presided over by the circumciser, an elderly woman whose eyesight can't be 20/20 and who in fact has the blue cataracts of age. She was dressed neatly in a white and red checkered dress and held her "crown" of authority, a stick of wood with a silver band around its middle, which looked remarkably like a penis. How ancient is this "crown," I wondered, and does it bear any relationship to the wedding finger and wedding band? I wondered this because the primary purpose of circumcision is to make the girls marriageable.

She pontificated at length about "the tradition" and how she was chosen by the village to perform circumcisions— gold, she said, was "poured" over her. Two bulls were killed. (This should have told her something, I thought, since the bull is a very ancient symbol of woman, the shape of its head and horns the shape of her internal reproductive organs and a symbol for them.) All the women before her had been circumcised. All those after her would be, also. She said this flatly, defiantly, looking directly at me. It was uncanny, this sense I had that she was aware, laying eyes on me, that we were implacably at odds. Soon a "man" came forward (so male-looking no one would dream he was anything else) and snatched up the white chicken that lay, bound

*The beheaded chicken and the feet of the mutilated girls. Dar Salamay, The Gambia.*

and trembling, at the base of the tree. Putting his foot on it to hold it still, he took a knife and hacked off its head. Blood splattered, some of it on little Mary's feet. Her feet, the smallest of all, wrecked me. I thought of the circumciser grabbing her and of her, her eyes taped shut, not even knowing what or who was grabbing her or what was sought. I finally started to weep, looking at those small feet.

Nobody else cried. They were laughing. The children were crying inside. I understood the message of the sacrifice: Next time, we cut off your head.

Later, interviewing the circumciser, I asked what she felt when the children cried and screamed. She didn't hear them, she said.

Interviewing her was very difficult. I glanced at her hands—extremely dirty, with black gunk under the nails—and thought of their coarse hardness against the tenderest parts of these girls. What do your assistants do? I asked. She could not say. Only that even without assistance she could circumcise the girls, because they'd lie perfectly still (I thought of the bound, trembling chicken) while she cut them. I asked what was in her circumcising kit, but she said it was a secret. I asked what kind of circumcision she did on the children. That, too, she said was a secret. I could not resist telling her it was not a secret to us. I described the various forms of circumcision: Sunna, intermediate, and pharaonic. I pantomimed infibulation, the sewing shut of the vulva after all external sexual organs are removed. She seemed genuinely surprised to know I had this information. How many of the children had died? I asked. None, she said. But of course if the children died, their deaths would be attributed to other causes: to sorcery, for instance. We sat there in her front yard, on small wooden stools that wobbled when we moved, the sun so hot I could almost hear it sucking the moisture from our skins. The circumciser wore a face of saccharine sweetness. My head began to ache. What relationship did she have with the girls afterward? They respected her, she said. Was it not perhaps fear? I asked. She smiled.

She was wearing some of her "gold." Very cheap and white-looking, one pendant in the shape of Queen Nefertiti, who was said to have been circumcised. It was chilling to think that for many African women this ritual of circumcision is the only real link they have with their ancient African-Egyptian heritage.

Nearly overcome by this encounter—shocking the circumciser by telling her that many of the women around her (myself and the crew) were not circumcised, a condition she couldn't fathom, never having seen such a thing: she actually drew back from me as if I'd said I had a disease—I was next interviewing two of the girls. One was holding a candy bar. The other sat quite still and silent, trying to keep her new head covering from sliding off (from this day she will have to keep her head covered in public). Their eyes and lips were the eyes and lips of so many people I know and love. I felt I'd been stabbed through the heart. By now my head was aching in earnest. They'd been perfectly indoctrinated and programmed to say nothing they felt. Indeed, the circumciser had told me they wouldn't tell what had happened to them even if someone put a knife to their throat. (The beheaded chicken, I thought, was the perfect symbol for this threat.)

So. It was a "tradition." Their mother had had it done to her. Her mother. The elders and ancestors. Yes, of course they would do it to their own girls. I wanted to take them in my arms and fly away with them.

Deborah and I joined the long and dusty procession as it headed back into the village: the circumciser in front with her bundle on her head, the two of us bringing up the rear. All the other women with their heads covered, the two of us rebelliously dreadlocked. Much feigned merriment, clowning, energetic dancing. A little girl, five or so, suddenly appeared out of nowhere and took my hand. Just for an instant. I felt she knew I had come for her sake. She was the "one African child" (that maybe my work against genital mutilation will protect) of my dreams. However, I know I am too late to protect *her* and this awareness provokes an almost engulfing wave of futility.

### 3:00 A.M., STILL FEBRUARY 9
*(still my birthday back home in California)*

Wide, wide, wide awake! I woke up thinking of my beloved. Feeling blessed by love itself. My life is wonderful. Even witnessing all this pain, I am glad to be in life. Today I am forty-nine, close to the end of perhaps the most fulfilling decade of my life. This happiness so unexpected and yet prepared for, even prepared for carefully. To be in good health. To be in love. To be doing work that will mean greater health and happiness to many. To be doing the work of protecting our children. To be in Africa. To realize Africans are doing OK, basically, if they'd just stop hurting themselves. And that I love both Africa and Africans. That Africans have "time" and "space." Westerners no longer have that. Africans really should be able to be wise, not just clever or smart. It's amazing how clear the mind gets when there're no billboards to obstruct vision.

So. Happy birthday, my little wondrous brown body that has its period and is trying to get through (or begin) menopause—hence my insomnia! You have carried my spirit well. I honor you and love you and vow I will continue to care for you with all the love I have found waiting for myself in my heart.


### FEBRUARY 10, 1993

Soon I shall be going for a massage offered by the Scottish wife of our driver, Malign. Her name is Rose, she's small and blondish, and she loves The Gambia and all things (except harmful ones) Gambian. Besides playing Tracy Chapman's music everywhere we go, which is greatly in his favor,

*Part of the tree that was a birthday present to Alice from our driver, Malign. Banjul, The Gambia.*

Malign took me and Deborah to see a giant cottonsilk (ka-pok) tree, visually my most spectacular birthday present so far. It is in the middle of the street, too big, apparently, for anyone to dream of cutting down, and so they simply built around it. Unbelievably big, this tree. Its roots above ground so tall one could stand between them, as if in small rooms.

I spent a few hours today interviewing Bilaela, who is head of an organization that educates women away from harmful traditional practices. She's seriously overweight, though not, perhaps, by African standards, and coughs a good bit. She has a cold in her chest and is allergic to the dust. Dust is everywhere. Her quarters are small but adequate, on one of the many unpaved streets of Banjul. She explained that she was eighteen when circumcised and that she resented it. While she was away from home years later, her aunt circumcised her two daughters. They now work with her against the practice.

Pratibha says I'm naive because I wanted to give Bilaela extra money—beyond what we'd paid for the interview and other help. But I had no idea until she told me that she'd already paid Bilaela seven thousand dollars! This seems very adequate considering Bilaela's frequent changes of mind about how and when and whether to assist us. So I put my five hundred bucks back in my pocket.

All I want for my birthday, I said to Deborah and Malign, is wonderful music! So off they went in search of some. Also for eyewash. The dust is murder on my eyes. They succeeded in getting some great music by a local musician named Demba. His songs are sweet and light and bouncy, overflowing with optimism and youth. Charming.

Later Deborah and I went to the Lebanese restaurant for dinner. Excellent. Before that I had another massage by Rose. Wonderful. By the time she finished, I was a noodle. She's completely against genital mutilation and is worried that Malign's daughter by his Gambian former wife will have it done to her. She and Malign are building a house together, and she plans to take the child to live with them. While she is massaging me, she asks her assistant, a local woman, whether she has circumcised her daughters. The woman says yes, but that she had it done in hospital. She says "hospital" with such a sense of satisfaction that I groan. For many people, "in hospital" is the answer, when to me the only answer is not to mutilate.

Unfortunately the food at the hotel is horrible. The fish is fresh but seems to have been sautéed in crankcase oil. Besides, all the former colonials vacation here, and their attitudes are those of an earlier millennium. Deborah overheard one such person (a woman) say to the waiter: "Oh, *boy,* Master would like some water!" If I could lift the ocean, I'd dump it over "Master"'s head. I find it hard to look at them, or, when I do look, to see them. The Africans who work here are delighted to serve Deborah and me. They love our dreadlocks and linger around the pool, covertly watching us (and smiling) while we swim. I love watching Deborah swim myself. Her locks are long and thick and stream out behind her magnificently as she moves powerfully through the blue water. Who knows, maybe we're the first black people, or black women, they've ever seen in this pool. We tried swimming in the ocean but were driven back to the hotel by the hordes of intrusive men on the beach.

On the day we arrived, I went out to the ocean to say a prayer to our ancestors who left this coast in chains, bound for the Americas so long ago. Within minutes I was set upon by local men, who treat women as if they have no other role in life but to respond to any inane comment any man makes to them. Finally I said, "I'm praying, do you mind?" This seemed to stun them, but because they're Muslims, they have to at least pretend to respect prayer. However, as soon as I left my prayer spot they were after me again. What was my name? Where was I from? Didn't I need to buy this or that? Would I come to this or that restaurant (pointing to a leaning table beneath a wilted palm in the far distance)?

Somehow I don't think tourism will thrive in The Gambia unless tourists have a thicker skin than mine. There's no such thing as a woman having a quiet moment on the beach alone. Or anywhere else, for that matter. Women are routinely followed, yelled at, harassed on the street. I can't help but connect this behavior to genital mutilation: the acceptance of domination, the lack of a strong sense of self one sees among the women here. Or, conversely, there will occasionally be an extremely loud, brash woman, like the one who pressed us to buy her wares with such vigor that she ran us out of her stall. These are the women whose pent-up anger seems to be a powder keg.

FEBRUARY 11, 1993. *Pratibha's Birthday*

There are four Aquarians on this trip: me, Deborah, Pratibha, and Nazila, our production manager. We are, our group, racially and ethnically mixed. Quite beautiful, I think, to behold. Nancy, the camerawoman, is the only white person. She's also the tallest. Then there's Nazila, who is Iranian. Deborah and I, both African-AmerIndian. Pratibha, Indian, born in Africa, living in London. There's Judy, whose parents were born in Jamaica and who now lives in London, and Lorraine, who comes, by way of her ancestors, from, I think, Trinidad. Because there are so many Aquarians, we've decided to have a communal birthday party at a nearby Gambian restaurant. All morning and much of last night, I've worked on my present for Pratibha. A poem. Because she's Indian, a bit of the tone of the Bhagavad Gita creeps in, tickling me, and so I'll leave it.

## Poem for Pratibha on her 38th Birthday

FEBRUARY 11, 1993.   THE GAMBIA, WEST AFRICA

Everything and
everyone
to the girl child
of Africa
appears to be
against her.
This was the message of
the dream
I had last night
in which I stood
with hammer raised
above the head
of a white colonialist
reciting his crimes
and an African girl
child
my youngest self
now
held the gun.

It was a healing
dream
nonetheless
for this much
I at least
see
and feel
clearly:

As I always
suspected
She
the very foundation
of life
and all that is
lovely
is deliberately
spoiled
ruined
abused
tamed
made into an evil
scowling
woman
who encounters
the world
with split mind
and filed teeth.

O Pratibha
we have come to
this obscure
place
under a bright
moon
bringing
our feeling
hearts
intelligence
and our

willing and
capable
hands.

The knowledge of
the extent
of our
foundation's
destruction
was hidden
from us
cleverly
hidden
by those
who knew.

The African mothers
forced to "forget"
their pain.

The African fathers
trapped
at last
into a shamed
and frustrated
secret
brotherhood
the bond
being
their
daughters'

mutilated
flesh.

The white
colonialists
who used
this cruelty
to children, an
African cruelty,
as
justification
for their own
savagery
against
the people.

The list is
long.

Against this
you place
your small
body
brown as
earth
and your
dark
eyes
stunned
with
hurt

and your
wise
hands.

As we work
together
we begin
to rebuild
the
shattered
ancient
foundation
of
the
universal
family
of women.

And so
on the day
of your
birth
let the
oceans
roll
for you
let the
sun
shine her light
on you
let the

wind
play in
your hair
that is so
like the
wings
of
blackbirds
let
the dust
of
earth
embrace
your feet.

Now is
the time
for
love's
warriors
to come
to
her
rescue.

Not with
guns
O
Pratibha
but with
every word

of
encouragement
we know.

With
every smile
of hope
that can
be snatched
from the
frowns
and wails
of our
belated
understanding.

With
every touch
& each
embrace.

The true
energy
of earth
and that
of
all life
is love.

By our
presence

here
in this
shadowed place
under a
scorching sun
and
cooling bright
moon
we return
love's
energy
to one
long
missing
from its
healing
circle
without
whom
of course
there has
been
no circle
and love
itself
has
languished
in
corners
of the
heart

never
in its
fullness.

On this
day
O
Pratibha
we bless
you
and call
on
all the
ancestors
to do the
same
and all
the powers
from
the four
directions.

On this
journey
I see
the African
child
with new eyes
of
happiness
and

delight
for I
know
I would
lay down
my life
for her.

O
Pratibha
Sister
and Love's
warrior:
May
the tide turn
and
the new day
break
for us
all.

We found ourselves in a nearly empty Gambian restaurant run by warm, shyly smiling Gambians. I read Pratibha's poem to her, and both of us were teary-eyed. Then we feasted on delicious Gambian food, which reminded me of the best New Orleans cooking in the United States.

Later that night, listening to the lapping of the waves beneath my window, I composed a second poem, for my beloved, so far away:

### Missing You

The waves of the sea
are not more eager
to reach the shore
than I am
to reach you.

The ocean
is in my bed.

FEBRUARY 12, 1993

Little sleep last night. Maybe the tea I drank—I can't tolerate black teas—maybe menopausal insomnia. Maybe missing my beloved. Today is Friday; tomorrow morning, quite early, we leave for Senegal, overland, by bus. On Wednesday we left the hotel and set out for the town of Kerewan, taking two ferries across the enormous Gambia

River and one of its tributaries. Very beautiful crossing so early in the morning. Cool, quiet, and gray, with a silvery cast to the water from a very weak morning sun. A bit like being inside a giant pearl. Unfortunately the car in which Bilaela and her daughters were traveling overturned, and though no one was seriously hurt, it changed the feeling of the day as we went about searching for clinics and a doctor. We had planned three exciting events. One of them a traditional naming ceremony, and one a traditional wedding. Exciting for us, perhaps, but what about for the bride? In many traditional weddings the bride ends up on her knees with a shroud thrown over her by her new master. Bilaela, though obviously disoriented and in some discomfort, insisted on taking us to see a garden project run by women to whom she had given the land when she had an opportunity as head of some agency or other (the third event!). These were happy, proud women! And their gardens were spectacular: onions, tomatoes, eggplant, peppers, herbs. Mango and papaya trees. Many deep wells. An oasis in an otherwise hot and dusty landscape. Deborah and I, the only gardeners among our group, were instantly taken up by the women and walked over the entire huge garden. We oohed and aahed aplenty, and quite sincerely, because what they've done is amazing. If only they had some way of moving their produce, they could easily feed all the people around them, marketing the surplus in villages and towns. Alas, there's no such transportation. It was all women, with the exception of one very sweet young man, nearly nude, deep in the earth, digging a well with a pick and shovel, his girlfriend sitting at the rim and hauling up the buckets of dirt. Deborah and I were drawn to them, and though we couldn't speak each other's languages, we let them know we, too, appreciate the miracle of water flowing out of the earth and the transformation of desert to garden.

FEBRUARY 14, 1993. *Valentine's Day*.
POPENGUINE, SENEGAL

At last I am in Popenguine, in Ayi Kwei's welcoming house. All love and affection. Friends. Julius Coles (Ayi Kwei's neighbor, amazingly enough, who has a small Moorish-style house just across the road) just walked up the drive and presented me with a bottle of Vouvray, which we will all certainly enjoy with dinner. He was all smiles and gentleness, reminding me of our long friendship, since he was a senior at Morehouse and I a freshperson at Spelman. We've managed to bump into many of each other's journeys, and so it seems right that he should be here, in West Africa, offering to introduce me to all the people he knows who can help us make our film. He suggests an African-American woman from the United States, married to an African, who has made a business of helping foreign film crews. I am delighted to have her number and will pass it on to Pratibha and the crew.

We arrived yesterday in Sindia, a small roadside village, on our long, bumpy, hot, and dusty ride, via bus, from Banjul. A five-hour trip, with a couple of stops at out-of-the-way villages and a failed meeting with a circumciser, who, at the last minute, decided not to appear. Ayi Kwei, whom I sometimes call "born and beautiful" because of his book *The Beautiful Ones Are Not Yet Born,* met us by the side of the road and brought us via taxi to Popenguine and

*Alice and Ayi Kwei meet by the road in Sindia, Senegal.*

‖‖‖‖‖‖‖‖

his home, with its peaceful view of the sea. We were delighted to set eyes on each other again; it has been nearly two years since he was in Berkeley at the university and I got to know him and his children, to whom I am now aunt. Into his tender black arms I went, my grin stretching from ear to ear. Later, after a walk on the beach and delicious fresh fish, we talked for hours, even though I was very tired; Ayi Kwei's beautiful voice sounding rather like the ocean, a low rumble in my ear. Eventually we went off to sleep. It is so dry and dusty here I had trouble sleeping, though not as much as I'd feared. I awoke only a couple of times, feeling thirsty.

At some point Deborah and I also went for a walk on the beach, and we fell asleep in the shade of large boulders.

Delicious. And when we woke up, her stomach was better (changes in food have been hard on us), I was more content (able to let go of some of the scenes I'd witnessed), and the wind and dust didn't bother me quite as much. For the wind is constantly blowing. The dust always in nose and throat. Midday looks like dusk, though it is of course still very hot.

My beloved called to wish me Happy Valentine's Day, and it feels odd to think of a Valentine's Day kind of love here, where men have three and four wives, all of them poor, and even the mosque is forbidden to women until after menopause, when they are considered closer to being male. We strolled past the mosque one day, and it looked totally dead and boring. A few old men on their knees; mechanical prayers blaring from an ancient loudspeaker in a language many of the people do not speak. Senegal is dry and dusty and poor beyond my imaginings. What has happened to these people, that they seem so joyless and oppressed? Is it Islam, as some suggest, which encourages passivity and desertification? Everything, including massive overgrazing of livestock, turning the fertile land to desert, is merely "the will of Allah"? I think genital mutilation plays a role. The early submission by force that is the hallmark of mutilation. The feeling of being overpowered and thoroughly dominated by those you are duty bound to respect. The result is women with downcast eyes and stiff backs and necks (they are of course beaten by fathers and brothers and husbands). And men who look at a woman's body as if it is a meal. On the other hand, there are all the Africans, women and men, I'm finding to love. People like Malign and Ayi Kwei, who are so tender, sincere, and *true* that I just spontaneously hug them. Although Ayi Kwei is rather formal, and so I have to constantly ask him: Am I too affectionate with you? He smiles, like the sage he is, and says, wisely, No-o-o.

My hair is dry. The sun just fries it. My skin is better since I rubbed myself all over with massage oil. Ayi Kwei often comments on my energy, even when I feel I'm barely moving. But I do have energy for what I'm doing, and I'm very grateful for that.


TUESDAY, FEBRUARY 16, 1993

My memoirs should be called "In Bed" because so many of the entries begin that way. So:

In bed in the Savana Hotel in Dakar. In a suite right on the ocean. Bougainvillea and palms. Very pretty, and comfortable as well. Popenguine was all I could have wished for and more. A lovely if very poor village. Yesterday Deborah, Asar (Ayi Kwei's eleven-year-old son and my nephew), Ayi Kwei, and I walked along the beach to the school, where I was questioned by the children. Such bright, thoughtful faces! The school itself is pretty, yellow cottages with blue trim. It didn't surprise me somehow that years ago, in 1961, when I was just getting to know Julius Coles, he was with a group called Crossroads to Africa and actually helped build this very same school. There's a simple plaque commemorating that fact. It partly explains his love of the village and his sense of belonging. The day we arrived, we encountered him fishing on the beach, as if, hundreds of years ago, he'd never left. Ayi Kwei and his friend the teacher translated for me. No talk about genital mutilation, because the headmaster already looked a bit nervous. Instead, I was asked about meditation! So I gave a demonstration. Fun. I explained to the children that when I teach, I always begin class with ten minutes of meditation. One of their teachers, Muslim, as they all are, attempted to explain meditation to the children as an occult practice and therefore something suspect. But I told them it was simply a way of attaining a clear mind and clear thinking. There is one woman teacher at the school, recently hired, looking like a butterfly—because of her brightly colored boubou— among moths. But very timid and expressionless. The women here chew sticks with which they clean their teeth. They chew them all day long. She had one in her mouth, sticking out like a cheroot. It is a bit like seeing someone all dressed up and walking about in public, but with a toothbrush clenched between her teeth. But what if there is another purpose? Something to bite down on instead of one's tongue?

Later Ayi Kwei said: You were very good with the children; you didn't condescend to them. I told him that actually I used to be afraid of children, they'd caused me so much grief as a child. But now I am able to see them more clearly and one thing I've vowed never to do is lie to a child. For that, to me, is the root of all evil. While at my house in Mendocino one day, we'd talked briefly about genital mutilation. Genitally mutilated women he's known have been very angry. I think now about what that means in a woman's relationship to her child. Does it mean she's often abrupt, cold, withholding, abusive? Or simply that she never smiles, which might be the greatest abuse of all?

Ayi Kwei and Mas (Asar) and the taxi driver brought us

into Dakar, the outskirts of which seemed unbearably poor and dusty and depressing. In fact, the road from Banjul to Dakar is so bad there are long stretches where motorists simply drive along beside it, not wanting to damage their tires on the bumps. At the Hotel Independence (named after some long-forgotten shining moment), we were informed that the crew had departed and gone to the Southern Cross Hotel. Off we went, too. It was horrible. My closet-like room overlooked a yard filled with rubbish. I called Julius, and he and his friend Alice rescued us. Hence my suite overlooking the ocean and Deborah's room with its view of Gorée Island. We had dinner at Julius's palatial house and listened to his tape of Bobby Blue Bland, of all people! He is joyful, enthusiastic, in love, and always trying to help— and succeeding. It is the best feeling in the world to travel so far from home and to find home wherever you are.

However, the Dakar he knows is not the one Ayi Kwei knows or the one I've been seeing. Julius's Dakar is rich, and the air isn't even dusty. In fact, it's the old colonial city. Ayi Kwei's Dakar is poor, dusty, dirty, noisy, and it smells. The stench comes from the absolute inadequacy of the drains, which overflow into the street. While fasting, devout Muslims are always spitting, adding to the discomfort one feels, walking about.

Julius took us to dinner again last night, stopping along the coast to show us how beautiful Dakar is. He is in love with the city, the ocean, the sky above his head. Later we had a rendezvous with members of the crew who wanted us to experience Le Ponty, a place of prostitution. It looked like a very sleazy sandwich shop, and Deborah and I left after a few minutes. The thought of prostitution is always horrible, but the thought of genitally mutilated prostitutes was more than I could tolerate. Besides, my malaria pills make me tired and drowsy. They affect my sleep, my vision, and my equilibrium. A real drag.

A meeting yesterday with seventy to a hundred women in a spacious courtyard in Dakar. Most of the women sitting on the ground, where Deborah and I joined them. Awa Thiam, who has written so powerfully on the status of African women, was present. I have looked forward to meeting her for many years. She has a heroic look, to match her work. She is tall, slender, beautiful, reserved. These days a politician, very involved in the present election campaign. Which, ironically, Julius is also involved in, to the extent of authorizing USAID money to be used in increasing voter participation. Voters must be apathetic, I think, if they have to be jogged from outside. These city women are quite cool and cynical, I thought—until we encouraged their offer to dance. Up to the dancing, we were trying to hold a conversation—about wife-beating, child abuse, economic empowerment—in French, Wolof, Mandinka, and English. In the middle of this I was stunned to hear Madame Fall, our host, ask me to buy them a refrigerated truck. Having seen some of the splendid gardens the women have, I realize this is just exactly what they need to get their produce to market; still, I assured her that a whole refrigerated truck is a bit out of my range and the film's budget. Maybe I could contribute a couple of tires? The dance the women did for us was exaggeratedly erotic. Several women danced, sticking out their tongues and with their eyes rolled back, presumably in ecstasy. So curious to see in a mutilating culture, and so repetitive and learned-looking. When I danced I felt very stationary by comparison, for I danced the way we do at home when it's too hot to move very much, slowly and suggestively, the effort all internal.

The last two days of shooting occurred on Gorée Island. The crew and Pratibha and I took the ferry early in the morning and walked immediately to the House of Slaves, a large Moorish structure of reddish adobe. Pratibha filmed an interview with me there, as I tried to put words to the feeling of sorrow I felt when I considered the possibility that not only had our maternal ancestors suffered all the horrors of enslavement, of which we mostly know, but they might have been mutilated genitally as well. The thought of these ancestors crossing the ocean in the holds of ships, in chains, with their vulvas stitched almost completely shut, nearly drove me round the bend.

Joseph (Ndiaya), who is administrator of the House of Slaves, looks just like my mother's brothers. I went in to thank him for assisting us in our filming. He said there was nothing to thank him for. He said, Perhaps you and I are cousins. We were sitting close together, because someone had brought a low stool for me, which I placed near his desk. The fatherly kindness of his voice and his look of being a relative completely undid me, and I started to weep, completely without intending to or even thinking that I might. Deborah, standing near the desk, began to weep also, and I suddenly realized that this is probably the response of many African Americans who come to Gorée Island, their ancestors' last bit of Africa before being shipped to the Americas. Earlier I had interviewed Tracy Chapman, who was visiting Gorée Island for the first time, and saw she had been quite shaken by the experience. She was amazed when I told her

*Alice's arrival on Gorée Island, on the way to the House of Slaves.*

|||||||||||||

*Alice's statement on Gorée Island.*

|||||||||||||

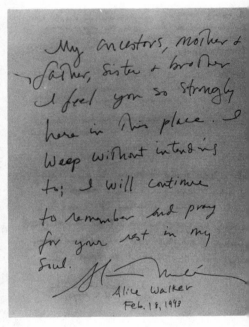

My ancestors, mother & father, sister & brother I feel you so strongly here in this place. I weep without intending to; I will continue to remember and pray for your rest in my soul.

Alice Walker
Feb. 18, 1993

**B. Joseph Ndiaya,** *curator of the House of Slaves Museum, presents to Alice a sculpture of a black man in chains.*

|||||||||||||

*Alice and Deborah at the House of Slaves.*

||||||||||||

*A sign showing when slavery was abolished in these countries. It is chilling to see that slavery was officially ended in Africa a hundred years after it was abolished in the West. It is still reported to occur in Mauritania.*

||||||||||||

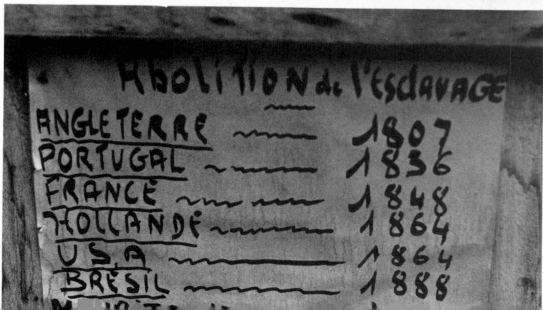

Abolition de l'esclavage

| ANGLETERRE | 1807 |
| PORTUGAL | 1836 |
| FRANCE | 1848 |
| HOLLANDE | 1864 |
| USA | 1864 |
| BRESIL | 1888 |

I had a gift for her from someone in The Gambia who loved her music, and I was able, later, in Dakar, to give it to her.


FEBRUARY 22, 1993

Though Pratibha and the crew returned home to London and the United States after our shoot on Gorée Island, a friend from the U.S. joined me for the last leg of the journey. To Burkina Faso, formerly, under the French, Upper Volta, whose new name means "Land of Honest People" and was given it by its Marxist president, Thomas Sankara. He was assassinated in October of 1987, many believe by the present president of the country. He was thirty-seven years old, a brave, compassionate, and brilliant leader, in the style of Patrice Lumumba.* Before his death, he made one of the strongest statements against female genital mutilation of any African leader. It is for this reason I make a pilgrimage to his grave.

It is the first thing I do in Ouagadougou.

The people have been forbidden to place flowers or any other token of remembrance on his grave. (The story goes that it was even forbidden to bury him and that his wife, like Antigone, covered his body with sand.) But those who loved and respected him do so still and have left many small piles of stones. There are many such piles in the shape of the cross, for this country, unlike Senegal, has a large Christian population. It is not the cross that has meaning for me but the circle. I make a circle of stones above where

---

*Patrice Lumumba, the first indigenous revolutionary president of the Congo. He was assassinated in 1961.

*Thomas Sankara, the young president of Burkina Faso, who was vocal in his opposition to female genital mutilation. He was assassinated in 1987.*

||||||||||||||

*Alice placing a circle of stones on Thomas Sankara's grave. Ouagadougou, Burkina Faso.*

||||||||||||||

I assume his head to be. I pray for his spirit to be at rest, after what was done to him (and to his comrades, as well; there are many graves around his of men who died the same way, the same day). I thank him for having had the courage and understanding to join women and children in their fight.

(Many months later, in Havana, where I am with a delegation delivering medicines to Cuban children and challenging the U.S. embargo that is rapidly undermining their splendid health, I speak about genital mutilation with Cuba's president, Fidel Castro Ruz. He immediately grasps the seriousness of the issue and says I must get all the leaders, the educators, the medical profession, and the politicians involved in educating people away from something that clearly harms them. I tell him about my book *Possessing the Secret of Joy,* a copy of which I've given him, and about our film. When he expresses interest, I promise a screening of it in Havana. He is emphatic and passionate in his concern. Shocked. He says he's never heard of this before and has innumerable questions, from the health consequences for women and little girls to who is putting up the money for our film. I miss Thomas Sankara keenly then, the African leader whose response most approximates Fidel Castro Ruz's. I grieve anew over what, by his assassination, we as women, as children, as Africans, as a world, lost. The men who stand with women and children can expect to suffer, and yet, for the earth to be saved, this is exactly what they must do.)

A second reason for going to Burkina Faso was to attend the all-African film festival, which is held in Ouagadougou every other year. I was curious to know whether this might be an appropriate African forum for our film.

For although genital mutilation occurs in London, Paris, and the U.S., it is in Africa (and parts of Asia) that it is so routine that a woman is, in many cultures, considered a monster if she is not "circumcised."

The festival is primarily a male affair, although recently women filmmakers have presented their works. My friend and I saw five films in as many days, using the rest of the time to explore the city of Ouagadougou and the countryside surrounding it. We saw an excellent film about Patrice Lumumba. He was so grand and beautiful to see, even after Mobutu had captured and beaten him, that my heart filled with a love that could not be drowned, even by my tears. What was Lumumba's crime, after all? That he loved his people, that he wanted the wealth of the Congo (now Zaire: they changed the name because the other had such bloody associations) to benefit the Congolese, not the colonizing French. There was a very unsatisfying film about Thomas Sankara. The day it showed, the theater was surrounded by soldiers. The place was packed. There was an announcement that the film had not arrived. People were unhappy to hear this. Eventually, after two films about the unfamiliar courting customs of a Berber tribe in a remote corner of northern Algeria, it was shown. The filmmaker did his best, but the final words of the film are those of the present president, Sankara's alleged murderer, and the assassination is not dealt with at all. It was still wonderful to see Thomas Sankara, so vibrant and confident, and his mother and father and wife. And to hear his voice, fearless and strong. And well-educated. What would happen if every well-educated person put his or her skills to the service of all? When you lose someone like Sankara, as when Egypt loses someone like Nawal El Sadaawi, or we lose Medgar Evers, Malcolm X, and Martin Luther King, or Bolivia (and Argentina) loses Che, or Grenada loses Maurice Bishop, or Mozambique loses Samora Machel, it sets the country and the people back a hundred years, if not more. *But only if the people agree* to be set back. Maybe we should learn to accept the deaths of our beloved leaders as an opportunity to develop our own self–leadership capacities.

The people of Burkina Faso have dignity and self-possession. Because it is a poor country, everyone rides a bicycle or a motorbike. This gives Ouagadougou a serene, timeless air; and to their credit, the French planted lots of trees, which now shade the wide boulevards. Still, I sense the collective grief just beneath the surface, exactly like the grief with which people in the U.S. have tried to live, with varying degrees of success and failure, these past thirty-odd years, beginning with the assassinations of Medgar Evers and John F. Kennedy.

There was a very moving documentary by a woman filmmaker about Mozambican women refugees in Tanzania who crush rock, using lead pipes, for a living. There was one film about genital mutilation, but we were never able to find the theater in which it was shown.

One problem I anticipate is that our film is in English. Most films in the festival are in French, and there are no subtitles. Still, I believe it will be worth trying to show the film here. Students, filmmakers, intellectuals, and artists come from all over the continent to view films at this festival. There will also be the possibility of showing the film at African universities. While planning the film, I dreamed of taking it from village to village, but by now I've visited many African villages, and there are absolutely no audio-visual facilities. Barely, sometimes, drinking water. None that we foreigners could drink.

It is the poverty of Africa that stays with me, along with the bright spirits of African children. The exhaustion of the land, especially in Senegal, and the poor health of so many of its people. Léopold Senghor, poet and former president of Senegal, now lives in France, I'm told. Many other African presidents appear to be trying to create Europe where they are, for example Houphouet-Boigny's replication of the Roman basilica in the Ivory Coast. Much of this is understandable, this longing to be where there is comfort, and plenty, and "freedom." But no cathedral or basilica, not even (or especially) the basilica in Rome itself, is worth the suffering of a single child. Africa and the world must choose.

Little did I suspect, when I was congratulating myself for being happy in the midst of so much pain, that I would begin weeping on viewing the rushes of our film in London and would continue weeping for two months. I've been crying over the lives of women and children who've been dead, some of them, thousands of years (one estimate is that genital mutilation is at least six thousand years old), as well as mourning the lives of those I've encountered, whose suffering is apparent. I grieve, as well, for all those children whose future holds a day of seizure, of being pinned down by adults known and unknown, and brutally hurt. And I accompany some of them, in imagination, through their efforts to forget, overcome, endure, even forgive a frightening and unforgivable injury. At various points I considered taking Contac or some other medication in an attempt to dry up my tears. Eventually, when I reached the bottom of what seemed more like an ocean than a well, I began to weep less. It was when I began to think of all those mothers and older sisters who must have fought for themselves and for their daughters and younger sisters over the millennia. How many of them must have been beaten into insensibility. Into the very earth. Chased out of the village, to die among the "wild beasts." How many must have been murdered outright. Sold into slavery. How many made to feel "crazy" and unclean. I began to feel strongly the power of my connection to their spirit of resistance and to find yet another honored and sacred place waiting just for them, for us, in my heart.

*A day without a date.* THE GAMBIA, WEST AFRICA, 1993

Today we journeyed in search of termite hills. They are all over the place, or what remains of them are. The farmers (frequently women) sometimes tear them down, presumably feasting on the termites, fried in oil, and use the material in the construction of their own dwellings. They crush the heavy, concrete-like substance, which must be earth mixed with the saliva of the termite, and add that to their own adobe (mud and straw) to make bricks. This produces an exceptionally strong brick and an enduring house. Anyway, we drove and drove and drove some more down many a dusty road, until we finally spotted a lone termite hill, some fifteen feet tall, rising from a field near the road. We clambered out of the van, and Pratibha directed me to walk around it and to touch it. There seemed to be the slightest

*Pratibha, Malign, and Nazila "climbing the clitoris," i.e., a termite hill.*

humming coming from inside it, and I found myself feeling quite protective, as I imagined what was going on inside. A complete universe in motion, a world, a neighborhood, a community, existed inside this curious structure, this high-rise apartment building in miniature. During what famine or long trek across barren terrain had people learned to eat termites? This part of Africa, the Sahel, so dusty, dry, and red, is like the outback of Australia. And I think of the witchetty grubs, large pale white worms, the aboriginal grandmothers showed me how to hunt and roast and eat, when I was in Australia, a year ago, laughing merrily as I tried to get out of it. They tasted like buttery scrambled eggs. Surprisingly tasty. No doubt fried termites would be similar, since the grub and the termite are, I believe, related. Isn't everything?

I thought of the termite "queen" inside the mound and marveled at the way her life—unknown, of course, to her—has been a model for the life of women in many African cultures. And how she, oblivious, and with her own problem of enslavement, could not care less.

It was exceedingly hot. So hot that Lorraine, helping to set up the shot, fainted.

While Pratibha and I had discussed making the film back home in California and I'd played Labi's "Something Inside So Strong" for her, I'd told her my vision for the end of the film. I want to be walking down a road with African women, I'd said; I want it to be clear that we are going on with life, and I want the audience to feel it should rise up and go on into life with us. I want this song played over us. Well. Shortly after Lorraine gained consciousness and just as we were about to return to the hotel after filming the termite hill, what should we see coming down the road toward us but a group of colorfully dressed, fast-stepping women. Pratibha stopped them, explained who we were, and told them she'd like to film them. They explained that they were on their way to a christening and feast and were late. Luckily Malign recognized some of the women, who were from his village, and promised to drive them to their destination after our shoot.

So there we all were, on a dusty, middle-of-nowhere road in the far outback of The Gambia, walking with a steady if hasty beat. The women were beautiful and full of humor. Even though we didn't speak each other's languages, we managed to laugh a lot. They all had expressive faces, and their grunts and groans and chuckles sounded so familiar I felt I understood everything. Who are these crazy people? they were asking each other. And why are they bothering us? Must not be from around here. Naw, couldn't be. Their heads are uncovered, and they're all wearing pants! And look at all the different colors they are and their different kinds of hair. Smell a little peculiar, too. And so on. It was like a dream, really. I knew them so well, I felt; indeed, I felt I knew the very road we were walking on. And somehow I also knew that, together, we'd get to the end of it. Perhaps it was our own christening to which we hurried along. Perhaps the new child being named today is us.

*Alice with women on their way to a christening, shown in the last sequence of the film.*

# PART TWO

# INTRODUCTION

WE EMBARK UPON multiple journeys in our lifetimes, walking along infinite paths. Sometimes one of these paths finds us following a golden thread that leads us to people who give our lives a precious and sacred texture.

*Warrior Marks,* the film, began its journey when Alice Walker came to me with a faith that stirred my soul to rise up in inspiration, to give voice to images and words for our shared vision that women be free from violence. We would use our imaginations to bring us one step closer to illuminating the beauty in women's lives.

Journeys have many beginnings.

I was born in Kenya, East Africa, and schooled in England, yet I was brought up to think of India as home. My experience as a daughter of migrant parents who have lived and worked on three continents has shaped me in fundamental ways.

My grandparents were "encouraged" by the British colonialists in India to go to Kenya. Kenya gained independence in 1963, and my family moved to England a few years later. We found our new home far from welcoming. England was so infused then with outdated notions of itself as the mother country for its ex–colonial subjects that it refused to look at its own image as a decaying nation, as an empire long dead. In 1967, when my family and many other Indian families were arriving, feelings against people of color were running high and Paki-bashing was a popular sport. It was in the school playground, at the age of eleven, that I first encountered myself as an undesirable alien. Anyone who has at any time been seen as marginal, peripheral, and "other," knows what it is to be defined by someone else's reality.

In the mid-1970s, I worked in the community with young South Asian women to create a series of four posters called "In Our Own Image." One poster, titled "Self-Defense Is No Offense," was a photograph of a young South Asian woman practicing self-defense, with a quote by a fourteen-year-old Pakistani girl: "If someone calls me a Paki I will go and kick their heads in." This series precipitated a heated discussion in educational institutions across London, with school authorities arguing that the defiant images depicted in the series were intended to incite violence. The debate was covered by the London media, and it was during this time that I began to understand the power of images and imagemaking, and to become interested in the media as an outlet for political struggle and resistance.

I came to the making of videos and films from a background of political and cultural activism, not from film or art school. I have been involved in feminist and antiracist organizing in Britain since the mid-1970s. In the early '80s, I became active in a group called the Organisation of Women of Asian and African Descent. I also began work for a feminist publisher, and have taught a course entitled "Women, Race, and Culture" at the University of London. Since then I have written many articles and essays on race and feminism and have coedited anthologies of writings by black and Third World women. My accumulated experience continues to feed and inform my work as a film-maker.

It was my interest in the work of women-of-color artists, writers, and political activists that led me to my first meeting with Alice Walker.

I had come to the U.S. in 1989 to attend the Robert Flaherty Film Seminar in upstate New York, where two of my early videos, *Emergence* and *Sari Red,* were being shown. A monthly magazine I'd occasionally written for in England asked me to go on to California from New York, to interview Alice about her novel *The Temple of My Familiar.* This was made possible because June Jordan, a good friend whom the writer Toni Morrison has called America's "premier black woman essayist," had just moved to California. June arranged for me to meet and interview Alice, who was her longtime friend.

It was a very special day. June and Angela Davis, to whom June had also introduced me, drove me from June's house in the Berkeley hills to Alice's home in San Francisco. So there I was, standing on Alice's porch, being introduced to her by June. It was an extraordinary moment as I stood there together with the three women, all of whom I greatly admired and respected.

I'd been nervous about doing the interview and therefore had prepared very well for it. I had read *The Temple of My Familiar* on the plane; the ten-hour flight from England had gone quickly as I became absorbed in the book.

Alice and I talked for an hour. I recall how the sun filtered into the beautiful room where we sat, and I kept wanting to look more closely at the amazing photographs and artwork on the walls. There was a large photograph of an elegant and proud black woman who I later learned was Septima Poinsette Clark, "one of the most effective yet unsung heroes of the civil rights movement," according to Brian Lanker in his book *I Dream a World*.

After our talk, Alice took me for a stroll in the park opposite her house. Later that afternoon, I remember, we stood by the sink in Alice's kitchen, eating a sweet, juicy melon, while we waited for June and Angela to take me back to Berkeley. I recall this simple sharing with great pleasure. As we were leaving, Alice gave me a gift. It was *I Dream a World,* whose cover featured the photograph of Septima Clark I had greatly admired earlier. Months later, I used photographs from this book as part of a set design for my film *A Place of Rage.*

Soon after my interview with Alice was published, I was pleasantly surprised to receive a card from her. She had liked the interview and invited me to tea the next time I was in California.

By this time I had begun work on *A Place of Rage,* a documentary about June Jordan and Angela Davis. This film was an exploration of African-American women's roles in the civil rights and black power movements.

When people of my community were being murdered on the streets of England in the 1970s because of their brown skins, I'd been prompted to search for ways of understanding this inexplicable violence. The American civil rights movement had inspired me and provided a framework from which I could begin to interpret and understand the many perplexing issues involving racism and oppression, and this new understanding gave me the confidence to be articulate and assertive in my work.

The role of women in these movements had not been given adequate attention, and in my modest way I attempted to correct this in *A Place of Rage.*

My effort to create filmic spaces where women of color can reach each other across the various diasporas had begun with my first video, *Emergence,* in which Palestinian, South Asian, African-American, and Chinese women speak through their art about the similarities in their histories of migration, exile, and enslavement. *A Place of Rage* and *Warrior Marks* continue this testimony.

In January 1990, I went back to California to begin research on *A Place of Rage.* Before going on to New York to do further film archival research, I spent the day with Alice and her daughter, Rebecca. They took me for a walk on Pacifica Beach. As we walked along the ocean, I asked Alice if she would be willing to be interviewed for the film I was making about June and Angela. I had read about Alice's involvement in the civil rights movement and had decided that her inclusion would be a great gift to the film. That day was really the beginning of my friendship with Alice.

In December 1991, I received the manuscript of *Possessing the Secret of Joy* from Alice, together with an invitation to collaborate with her on a documentary about female genital mutilation. I was honored that Alice wanted to work with me, and I embraced her invitation wholeheartedly.

I had known about female genital mutilation before I read *Possessing the Secret of Joy.* I was aware that it was practiced not only in Africa and the Middle East but also in other parts of Asia and among small communities of Bohra Muslims in India, Pakistan, and East Africa.

I was also keenly aware of the debate that had erupted at the 1985 UN Decade for Women conference in Nairobi, Kenya. Many African women had reacted angrily when Western feminists raised the subject of female genital mutilation. "Stop groping about in our panties" was their response, born of resentment at the colonial tone of Western feminism.

In the intervening years, African women have been neither silent nor inactive. They have organized conferences and built grass roots projects in a concerted effort to stop the practice of genital mutilation. Many African women have welcomed sisters from other countries who are willing to collaborate with them in spreading the word about their struggle.

Among the African women leading the crusade against female genital mutilation, the Senegalese Awa Thiam has been a leading spokeswoman for African women. Upon first meeting Awa, the author of *Black Sisters, Speak Out,* I asked her how she felt about our coming to Africa to make the film. Her reply was that what was happening to women in Africa should concern *all* women across the world, and this sentiment was repeatedly echoed by other African women we met on our journey.

During my work on *Warrior Marks,* I have had questions put to me by women from all cultures and races: "How do you feel about being an outsider to Africa and taking on this subject?" they have asked, and "Have you thought about questions of cultural transference and cultural imperialism?" often accompanied by such comments as "You are so courageous to take this on."

The fear of being labeled cultural imperialists and racists has made many women reluctant to say or do anything about female genital mutilation. Except for the writings and voices of a handful of white feminists over the last decade or so, there has been a deafening silence, a refusal to engage either critically or actively with this taboo area of feminist concern.

Clearly, female genital mutilation is a painful, complex, and difficult issue, which involves questions of cultural and national identities, sexuality, human rights, and the rights of women and girls to live safe and healthy lives. But this complexity is not an excuse to sit by and do nothing. Who cares if African women and children are subjected to violence? *We should all care.* If one hundred million white women and children were being mutilated as a matter of course in the name of tradition, the earth would by now be shaking with the tremors caused by voices of protest and righteous anger.

This reluctance to *interfere with other cultures* leaves African children at risk of mutilation. If we do not speak out, we collude in the perpetuation of this violence. There is no virtue in upholding, even unwittingly, the tradition of female genital mutilation.

I have asked myself certain questions that highlight the problems inherent in creating a necessarily complex and honest representation of this practice. These questions continued to inform my approach as I worked on this film. For example:

How can I create a sensitive and respectful representation of a people—and a continent—who have historically been grossly *misrepresented?*

How can *Warrior Marks* begin to challenge the cultural imperialist imagery of Africa and Africans, as perpetuated in Hollywood films like *Out of Africa?*

How can I work directly with African women in order to halt this violence against women and girls?

As a woman, I am concerned about eradicating the pervasive violence that all women experience across all cultures, all races, and all societies. The expression "Torture is not culture" tells us quite clearly that we cannot accept ritualized violence as an intrinsic part of any culture, or for that matter any sort of violence against women.

I feel enraged when I read about twenty-thousand Muslim women who have been imprisoned and repeatedly raped in the former Yugoslavia;

. . . when I read about women in India who, at a rate of one every two hours, are incinerated for not bringing in large enough dowries for their husbands;

. . . when I hear that girls in Asia are drowned and suffocated at birth because they are female;

. . . when I hear about girls as young as one month old who are bleeding to death because of excision;

. . . when I see girls as young as four being genitally mutilated in the name of patriarchal tradition and religion;

. . . when I hear about the death in 1992 of Hattie Cohn, an African-American lesbian woman whom neo-Nazis firebombed in Oregon;

. . . when I read about Joy Gardner, an illegal immigrant who was tied with a body belt resembling slave manacles and gagged with bandages by the British police, resulting in her death due to lack of oxygen to the brain.

For me, feminism is about creating a communal space in which women can attempt to change and abolish what is harmful to them, and it is these feminist impulses that have guided and fed my anger, rage, and spirit during the making of *Warrior Marks*.

Documentary filmmaking is far too often seen as the pursuit of objective truths. It is not. While making a documentary film is considered to be a quest for truth and reality, a dramatic film is thought to be an entirely imaginative construction. But in the end, documentaries are as subjective as dramatic narrative films. As a director of documentaries, I choose whom to interview, what to film, and how to put across certain images and words, and then I apply the same judgments I would in directing a fiction film with actors, sets, locations, and scripts. In both cases, it is my personal judgment and vision that inform the making of the film.

While filming in Africa and talking to African women about their initiatives and actions against female genital mutilation, we did not pretend that we were neutral or objective. I was open about the fact that I abhorred this practice, and I shared these feelings with the women I met and talked with. I recognized that I had come with my own judgments and political understandings of this practice.

I believe *Warrior Marks* is part of an ongoing project to speak out against the violence directed at women across the world. We need to be willing to transcend all our differences without ignoring them, to build new communities that bring us nearer to our utopian ideals, to continue to redefine our ideas about womanhood and feminist politics, and to embrace concepts of justice and equality, while at the same time recognizing the complexities of our diverse identities.

The future looks hopeful.

I see cause for hope when I read about mothers in Burkina Faso who are encouraging their daughters to fight against female genital mutilation. In their efforts to change attitudes and break down the resistance built up over the centuries, these women are pioneers.

I see cause for hope when I meet fathers and husbands who support the wishes of their daughters and wives not to be excised.

I see cause for hope when I see a growing awareness among women that genital mutilation is a pointless suffering rather than a necessary evil.

When women internationally join together in political unity

despite their differences, they become one of the greatest forces for change in the world.

I thank Alice and all the women who have spoken of their pain and shared their hearts and their lives so openly during the making of *Warrior Marks*.

In a world that is inundated and obsessed with images of inhumanity, I continue my journey to create images of profound humanity.

*Pratibha Parmar*
July 1993

# PRATIBHA'S JOURNEY

Dear Alice,

Happy New Year!

I was very excited to receive the manuscript of your new novel, *Possessing the Secret of Joy,* which I read over the Christmas break. It has been a long time since I have felt this incredible surge of emotions. It is an extremely powerful, haunting, and moving book. Truly formidable!

The character of Tashi remained with me for days; her journeying, both physical and psychic, was extraordinary and harrowing. I can understand what an emotionally exhausting time you must have had when writing and thinking about the book.

I would be very interested to work with you on a documentary about genital mutilation. As you know, I have wanted to work with you on a project since we met before I made *A Place of Rage.* I believe a complex and sensitive film focusing on this subject would serve as a necessary and important contribution to the struggle against genital mutilation. I am keen to make films which address women's experiences from an international perspective. (I have in fact been working with another Indian woman writer on developing a drama script around the issue of dowry deaths of women in India.)

What I find compelling about *Possessing the Secret of Joy* is the ways in which it brings together the complex and difficult issues of gender, sexuality, and "cultural" and national differences through the exploration of one woman's experience of being genitally mutilated. Difference is not neutral, and identity can be and is a necessary political resource, but both identity and cultural differences are constantly being reworked, transformed, and negotiated through the play of history, language, memory, and power. This central premise of the novel is challenging in the way it pushes the boundaries of current thinking on these questions.

One major contributing factor in the continuation of female genital mutilation is the lack of international pressure against this practice as well as the lack of educational resources being put into eradicating harmful traditions. It is important to see this as part of the continuum of violence against women across many cultures and societies. The form such violence takes varies, but ultimately what's clear is that misogyny is universal. The effects of patriarchal violence on the quality of women's daily lives are devastating.

In recent years, there have been documentaries on British television that have primarily concentrated on genital mutilation practices among the Somali communities in England. One of these films, called *Female Circumcision,* prompted the British government to pass a law in Parliament forbidding the practice within the British Isles.

I will begin to pursue potential funding for the documentary film that you are suggesting in your letter. However, I do still hope that the novel gets made into a dramatic feature film sometime in the future. Throughout the novel, there are so many rich visual metaphors, carried by a powerful sense of "poetic logic," as well as a mesmerizing narrative structure and memorable characterizations.

Anyway, these are some preliminary thoughts. There is much, much more I would like to say, but I will save it for when I come to California next month.

I feel honored that you have asked me to collaborate on this project with you. I know that together we can create a film that will be beautiful in its anger and hopeful in its vision, as well as useful in the fight against female genital mutilation.

I am looking forward to seeing you very much.

Lots of love,
*Pratibha*

*As may be a drop of water in the ocean, or yet a teardrop*
  *in the sea,*
WE MUST
*proclaim out loud what all women murmur in thought*
*denounce the crimes committed against women, mutilations*
    *that women*
*endure with resignation*
*offer resistance at every level*
    *active resistance*
    *effective resistance*
*to all oppression*
*whatever its source—unremitting*
*Only by countless voices in unison,*
*only by countless acts of resistance,*
*the countless sum of desires for change,*
*a limitless sum of goodwill*
    *will to change the nature of our lives, could we change the*
*present face of the world*
    *and straight away, the status of woman could end the*
*oppression and monstrous exploitation endured by women,*
*oppression and exploitation which were, and still are, in our*
*day, the daily lot of women. Strength will reside solely in the*
*multitude of voices, of people, of consciences resolved to*
*effect a radical change in all the present decadent social*
*structures. There is no other way.*

AWA THIAM

*There is no compromise between liberty and slavery.*

PATRICE LUMUMBA

*I will be a post-feminist when there is post-patriarchy.*

*Slogan on a T-shirt*

I read the first two quotes in the opening pages of Awa Thiam's essential text, *Black Sisters, Speak Out: Feminism and Oppression in Black Africa.* Of all the books I found during the research period of *Warrior Marks,* Awa's book left an indelible impression. This was the first book I had seen written from an African feminist position: radical, uncompromising, and devastating in its detailing of the oppression of African women. The chapter on clitoridectomy and infibulation contained direct and personal stories from women who had undergone sexual mutilation. The pain and the terror these women had experienced seeped through the pages, as did their anger and their rage.

One particular story, of a woman recounting her memory of being mutilated when she was twelve years old, was far more haunting and potent then anything I had read in the medical explanations of this practice or seen in the disturbing photographs of mutilated vaginas. I decided to use this story as one of the narratives within the film. Subsequently P. K.'s story became a central narrative, its emotional impact built in layers between the interviews. I have learned that stories that make the hair on the back of my neck stand up, that leave a gut-wrenching knot, are the ones that will have an enduring emotional impact.

### P. K.'s Story
*I had just turned twelve when I was excised. I still retain a very clear memory of the operation and of the ceremony associated with it. In my village, excision was only performed on two days of the week: Mondays and Thursdays. I don't know if this was based on custom. I was to be excised together with all the other girls of my age. Celebrations were held the previous evening. All the young and old people of the village gathered together and stuffed themselves with food. The tomtoms were beating loud, late into the night. Very early the next morning, as my mother was too easily*

*upset to have anything to do with the proceedings, my two favorite aunts took me to the hut where the excisor was waiting with some other younger women. The excisor was an old woman belonging to the blacksmiths' caste. Here, in Mali, it is usually women of this caste who practice ablation of the clitoris and infibulation.*

*On the threshold of the hut, my aunts exchanged the customary greetings and left me in the hands of the excisor. At that moment, I felt as if the earth was opening up under my feet. Apprehension? Fear of the unknown? I did not know what excision was, but on several occasions I had seen recently excised girls walking. I can tell you it was not a pretty sight. From the back you would have thought they were little bent old ladies who were trying to walk with a ruler balanced between their ankles and taking care not to let it fall. My elders had told me that excision was not a painful operation. It doesn't hurt, they repeatedly assured me. But the memory of the expression on the faces of the excised girls I had seen aroused my fears. Were not these older women simply trying to put my mind at rest and allay my anxieties?*

*Once I was inside the hut, the women began to sing my praises, to which I turned a deaf ear, as I was so overcome with terror. My throat was dry, and I was perspiring though it was early morning and not yet hot. "Lie down there," the excisor suddenly said to me, pointing to a mat stretched out on the ground. No sooner had I lain down than I felt my thin frail legs tightly grasped by heavy hands and pulled wide apart. I lifted my head. Two women on each side of me pinned me to the ground. My arms were also immobilized. Suddenly I felt some strange substance being spread over my genital organs. I only learned later that it was sand. It was supposed to facilitate the excision, it seems. The sensation I felt was most unpleasant. A hand had grasped a part of my genital organs. My heart seemed to miss a beat. I would have* given anything at that moment to be a thousand miles away; then a shooting pain brought me back to reality from my thoughts of flight. I was already being excised: first of all I underwent the ablation of the labia minora and then of the clitoris. The operation seemed to go on forever, as it had to be performed "to perfection." I was in the throes of endless agony, torn apart both physically and psychologically. It was the rule that girls of my age did not weep in this situation. I broke the rule. I reacted immediately with tears and screams of pain. I felt wet. I was bleeding. The blood flowed in torrents. Then they applied a mixture of butter and medicinal herbs which stopped the bleeding. Never had I felt such excruciating pain!

After this, the women let go their grasp, freeing my mutilated body. In the state I was in, I had no inclination to get up. But the voice of the excisor forced me to do so. "It's all over! You can stand up. You see, it wasn't so painful after all!" Two of the women in the hut helped me to my feet. Then they forced me, not only to walk back to join the other girls who had already been excised, but to dance with them. It was really asking too much of us. Nevertheless, all the girls were doing their best to dance. Encircled by young people and old, who had gathered for the occasion, I began to go through the motion of taking a few dance steps, as I was ordered to by the women in charge. I can't tell you what I felt at that moment. There was a burning sensation between my legs. Bathed in tears, I hopped about, rather than danced. I was what is known as a puny child. I felt exhausted, drained. As the supervising women who surrounded us goaded us on in this interminable, monstrous dance, I suddenly felt everything swimming around me. Then I knew nothing more. I had fainted. When I regained consciousness, I was lying in a hut, with several people around me.

Afterwards, the most terrible moments were when I had to

*defecate. It was a month before I was completely healed, as I continually had to scratch where the genital wound itched. When I was better, everyone mocked me, as I hadn't been brave, they said.*

Awa Thiam's book was written from her understanding of the universal pattern of patriarchal violence against women. Her belief in the power of international feminism to challenge and change was something that I found uplifting, especially at a time when so many Euro-American feminists are talking about postfeminism.

How can there be talk of postfeminism when more than three quarters of the world's women live on the poverty line? When female mortality rates are rapidly increasing? When it is estimated that by the end of the decade, 3.5 million women in Africa alone will be living with AIDS? To talk about postfeminism when girl children in Bangladesh, China, Afghanistan, and India are being regularly destroyed at birth, is to talk from a position of economic, social, and political privilege—which, incidentally, the majority of women even in Western countries do not enjoy.

My political understanding of female genital mutilation comes from my long-term involvement as a feminist reading and writing feminist theory and active in the diverse groups that have come to constitute the women's liberation movement since the 1970s in Britain.

As women of color, we have not absorbed acritically the Western feminist paradigm, which assumes a universal sisterhood where none exists. Over the decades, we have challenged the strategies and actions of white feminism that have ignored our racial and cultural differences and specificities. We have asserted our autonomy and built organizations that have given us forums for our voices.

As women of color, we have not accepted that we are marginal, other, exotic, dispensable, or in any way less valuable than any other human being in the world. The search for a womanhood, a selfhood, that takes into account our specific cultural and racial memories and histories has nowhere been more evident than in the proliferation of work by women-of-color artists, writers, poets, and, more recently, filmmakers, in both the U.S. and Britain. It is in the arena of cultural production that many of us have sought to give voice to our individual visions. And our insistence on incorporating an international perspective into our feminism makes us seek out like-minded sisters around the world.

Awa Thiam was based in Senegal, and it seemed clear, if only for that reason alone, that we *had* to go there. One of my fundamental guiding principles in thinking about this film was that it had to foreground African women's voices: voices of anger, analysis, resistance, and self-determination. An interview with Africa's first and leading feminist was absolutely essential.

Some of the documentaries that I'd seen on female genital mutilation tended to sensationalize the issue and show these practices as something outside the realm of Western civilization, something "other," "remote," "barbaric." In fact, the psychic and physical mutilations that women in the West undergo are equally devastating: unwanted hysterectomies, endless face-lifts, liposuction, bulimia, anorexia, silicone breast implants—all in the pursuit of youthfulness and an ever-changing notion of the ideal woman.

I began my research in earnest and read many books, including Hanny Lightfoot-Klein's *Prisoners of Ritual* and Fran Hosken's *The Hosken Report,* both of which were informative. I also reread *Possessing the Secret of Joy.*

Rita Wolf, an actress I befriended after her work on *Khush,* gave me Gloria Steinem's book *Outrageous Acts and Everyday Rebellions.* I'd liked the way, in the article on female genital mutilation she'd cowritten with Robin Morgan, Steinem had linked patriarchal practices like genital mutilation with examples of patriarchal oppression from the U.S. and Europe. I learned that Western nineteenth-century medical texts proclaimed genital mutilation as an accepted treatment for nymphomania, hysteria, and masturbation. Freud had a lot to answer for when he proclaimed: "The elimination of clitoral sexuality is a necessary precondition for the development of femininity."

The controlling, curbing, and problematizing of women's sexuality have always been cross-cultural. The similarities of certain myths and misconceptions from various continents and times are quite astounding. For instance, Freud's concept of "vagina dentata" has an almost exact counterpart in certain myths, folk tales, and legends recounted by Awa Thiam in her book. "The myth of the toothed vagina is found in many countries: for example, among the Bena-Lulua people in Congo, in Gran Chaco among the Toba, and among the Aino of Japan. It is surprising to note that the Bambara concept of the clitoris as a dagger is almost identical to that of the Toba, who view it as a residual tooth, presumably all that remains of the toothed vagina."

This ancient belief that if the clitoris is not excised it will grow razor-sharp teeth and engulf and eat up the penis is mirrored in Freud's concept. Different guises with the same purpose: to destroy women's right to autonomous sexuality in order to accommodate male fears and desires.

Having immersed myself in reading, I proceeded to write a treatment, which I sent in June together with Alice's statement about the film, to Alan Fountain at Channel 4 Television. Alan and I met in August. My original proposal was for a feature-length documentary, seventy-two minutes long, which would be interspersed with dramatizations from *Possessing the Secret of Joy*.

*Excerpts from treatment for* Warrior Marks, *submitted to Channel 4 Television. Format: 16mm; Length: 72 minutes*

### The film

*Warrior Marks* will be a film about journeys that overlap, collide, and create ruptures.

A journey that a writer makes both within herself and with her book.

A journey that the key character of the novel makes through her madness as she tries to overcome the trauma of her brutal childhood experience.

A journey that a filmmaker makes with both the writer and the character, Tashi, through California, Africa, and Europe.

### The journey of the writer and the book

I would like to film Alice Walker at her home in California before she sets off on a tour to launch her book. With what hopes, expectations, and thoughts does she set off? What does she think the reaction to her book is going to be? What preparations is she making within herself to face the potential hostility to her and to the book, for daring to speak about the "forbidden"? Why does she feel so passionately about this subject? How does she hope to change these age-old "traditions"? What research did she do for the book?

Subsequently we will travel with her and film the book launch in South Africa and Zimbabwe, capturing spontaneous reactions and responses to the book. Are the responses going to be any different in London? How does Alice feel about being in South Africa to launch the book?

### The journey of the key character of the novel, Tashi

We will dramatize some key elements of Tashi's journey through her madness as she tries desperately, through psychoanalysis, to regain the ability to recognize her own reality and to feel. As she studies the mythological "reasons" invented by her ancient ancestors for what was done to her and to millions of other women and girls over thousands of years, her understanding grows. But so does her grief and her anger, which propel her to act.

### The writer's journey to the sites where the ritual of female genital mutilation is practiced

I want to film Alice Walker's journey to Africa, more specifically to Kenya and Senegal. Alice would like to be filmed talking to elders, matriarchs, and young and older married couples about their sexual and psychological experiences of genital mutilation. She would like to visit hospitals where these "surgeries" are done.

### The filmmaker's journey with the writer, the book, and Tashi, the fictional character

I want to film the landscape of Africa with Alice Walker: remote villages where women clean themselves and their clothes by the riverbanks, where they collect firewood and carry it to their villages, men playing games with each other under the shade of a tree, women preparing meals, women preparing for their weddings, women giving birth, young women being prepared for the ritual of circumcision, faces of older women, women singing in celebration, women singing in mourning, the setting sun, the early-morning light, the details, the moods and the sensibilities of daily village life of the women of rural Africa.

I feel it is important to capture the essence of African daily life in the villages and contextualize the practice of female circumcision. The film will be removed from previous documentaries that have "looked at the issue of female circumcision." In this film, an ancient tradition will be exposed and discussed openly with women who have a diversity of feelings and opinions about it.

This will be a film that will explore three characters: Africa, the continent; Alice Walker, the writer; and the fictional character of Tashi. The film will move back and forth between the continents and capture the conflicts, the frissons, the tensions, and the drama.

Sequences of all three stories will be put together to create a nonlinear yet emotionally and visually textured film: an exploration of an ancient tradition through many eyes. The film will be a layered construction, a weaving of many images and sounds, characters and stories and landscapes, both fictional and real.

*26 August 1992*

Dear Alice,

Wonderful news! Last week I had a meeting with the commissioning editors at Channel 4 about our film and have just heard that they have agreed to give us some funding and a broadcast slot in their next autumn's schedule. I am so thrilled and delighted!

We had a good meeting discussing the film in detail. They are keen and enthusiastic. I had been worried about whether they would come through with the money, since there have been so many cuts in television budgets this year, but we did it!

The editors are Alan Fountain and Caroline Spry, who also commissioned some of my other films: *Khush; Double the Trouble, Twice the Fun;* and *A Place of Rage.* They said they were interested in working with me again and felt our proposal would make a good and challenging program. It's good to have their support.

They would like to show it on a series called *Critical Eye,* which goes out in the autumn of 1993. The time slot is one hour. This is a prime-time slot and usually attracts a large audience. *A Place of Rage* went out on this slot.

There are a number of points we need to discuss:

1. If we want to go to Sudan, Senegal, and Kenya, we will need to raise another $80,000, on top of what we already have. It is going to be difficult to raise this in England, and I wondered whether we should go to one of the foundations in the U.S.,

such as the Rockefeller or the Ford. I remember your mentioning this when I last saw you. Let me know what you think or if you have any other suggestions.

2. I have spoken to Sarah Wherry at Jonathan Cape in London, who has given me the schedule for your forthcoming visit to England. It would be interesting to film you at the Africa Centre with Efua Dorkenoo, whom I believe you are going to meet there on 14 October. I would also like to film the Institute of Contemporary Arts forum in which you are participating on Thursday, 15 October. These two events would be important to include in the film, as this footage will begin to give us a concrete sense of your journey with the book.

Let me know what you think, because I will have to start booking the crew and organizing things fairly quickly. We also need to set aside time to talk further about the film. I have booked with Sarah for you to come to dinner at our house on Saturday, 17 October.

3. I need to finalize the researcher/production assistant on this as soon as possible. Is Rebecca still interested, and is she going to be available for the times that we need her? I am enclosing a provisional schedule. If she is interested, could you please ask her to call me as soon as possible so I can discuss it with her further. I will be returning to California in October for a few weeks and am planning to spend December there. Meanwhile I send you many hugs and wait eagerly to hear from you.

Love,
*Pratibha*

While I was excited about getting funding from Channel 4, the drawback was that the film would have to be an hour long (which in television terms means fifty-two minutes, thirty seconds). Films made for television have to fit into particular time slots; feature-length documentaries are rarely shown. Even so, I was still delighted. The feature documentary would have to wait, but in the meantime I would grab the funding while it was available.

I also knew that in recession-hit Britain, many small companies had gone out of business, and many directors were out of work and struggling to keep going. In the television industry, it's hard to find financial support for independent films, especially documentaries that challenge the status quo. I felt lucky and somehow knew that this film was meant to be made. Now that the funding was in place, I could get moving and make some concrete plans.

I had to rethink my original proposal as a consequence of the financial restraints and the shorter time slot. I decided to scale it down by reluctantly removing the planned dramatizations from *Possessing the Secret of Joy*.

During this time I met Efua Dorkenoo, a founding member of the Foundation for Women's Health, Research and Development. (FORWARD). This group, which was later to become FORWARD International, is an independent organization established in 1983 to promote good health among African women, with a special emphasis on education against genital mutilation. Efua is a dedicated campaigner who has been working with African women at the grass roots level for many years, particularly in London. Campaigning against female genital mutilation has been a major focus of her work, and in fact Efua was the first person to introduce female genital mutilation as a human rights issue to the United Nations Commission on Human Rights in 1982.

Meeting Efua and listening to her countless stories of African sisters scarred for life by this brutality was both educational and galvanizing. Efua agreed to be our consultant. Her invaluable advice about which countries to go to in Africa, as well as her contacts, set us off on the right road.

## October 1992

*Possessing the Secret of Joy* was published in the autumn of 1992. Alice came over to Britain for the launch that her publishers had arranged. The book was originally going to be launched in South Africa, and I had hoped to go to film there, but the escalating violence meant that that trip had to be canceled. Instead, we planned a two-day shoot in London.

The core idea of the film I was working with was Alice's quest to explore the issue of female genital mutilation. I felt it was important that the film show Alice meeting different women whose lives have been affected deeply by this practice.

## Shoot Day 1
### 9 October 1992

We filmed Alice arriving and meeting Efua at the Africa Centre in Covent Garden. Since we couldn't find a quiet space for the interview and discussion between them, we filmed at the Ivy Restaurant, not far from the Centre.

Then we had to rush to the African Refugee Women's Group in Tottenham. This group, operating out of a small room in North London, provides counseling services for African women who are isolated and lonely, many of them newly arrived immigrants. I got a good sense of how vigilantly African women in the diaspora are working against genital mutilation, and of how this collective work has helped forge a sense of community among themselves. A Yoruba priestess sang a welcome song for Alice. Soon there was

*Alice and the Yoruba priestess, Tottenham.*

||||||||||||||

*Pratibha Parmar conversing with Alice Walker. River Thames, London.*

||||||||||||||

dancing, with Efua leading the way. We got some good footage of Alice and Efua dancing and laughing together.

Harriet Cox, a camera operator I'd worked with on previous films, did well shooting in that small space. She avoided imposing too much with the camera, yet captured some special moments. All day long we worked against time, because in the early evening Alice had to do a reading at the National Theatre on the South Bank, which was sold out.

*Shoot Day 2*
10 OCTOBER 1992

We arrived in the very early morning at the River Thames between Tower Bridge and London Bridge. Hardly anyone was there, which was unusual for a Sunday, a traditional market day. Maybe it's because it was a cold, overcast autumn day. The river didn't look too clean.

We filmed Alice with Aminata Diop. Alice had written to me about this courageous young woman, and suggested an interview with her and her attorney, Linda Weil-Curiel, for the film.

Aminata Diop had run away from her village in southern Mali in order to escape genital mutilation. This act of resistance cost Aminata her fiancé, her family, her future in Mali, and her way of life. She was twenty-two years old when she escaped to France with the help of a friend. Today she lives in Paris, and while the French government has granted her permission to remain there, it is not on the terms that Linda Weil-Curiel is fighting for. Her lawyer would like Aminata to be granted political-refugee status on the basis of gender under the Geneva Convention. "This would

*Alice Walker and Aminata Diop, London.*

‖‖‖‖‖‖‖

*From right: Pratibha Parmar directs Aminata Diop and Linda Weil-Curiel by the River Thames, London.*

‖‖‖‖‖‖‖

be the first acknowledgment of a woman's right to flee patriarchy," Linda says.

"Each rainy season in my village, they perform excision. Girls scream, suffer, are terrorized. One of my girlfriends was excised on a Thursday; by Sunday she was dead," said Aminata Diop in our filmed interview with her.

Our first meeting—Aminata, Linda, Alice, and myself—took place at my home, over a spicy Indian meal. Aminata, a shy young woman, was clearly happy to meet Alice. She had heard about Alice's book and knew that Alice was against female genital mutilation and supportive of her individual struggle.

Today, when we filmed the interview between Alice and Aminata, it was incredibly moving. Linda translated from French to English, and even with the language barrier there seemed to be a strong connection between Alice and Aminata. Everyone on the crew was affected when Aminata broke down as she struggled to articulate the pain of being rejected by her mother. Alice stroked Aminata's arm and held her hand, and finally she, too, broke down.

When Aminata said, "My mother says a woman has to go through three ordeals in life: We must go through excision, marriage, and giving birth. Excision is a woman's destiny," I wanted her to know that by her actions, she is bravely paving the way for her younger sisters to challenge the notion that it is a woman's destiny to be genitally mutilated.

Aminata misses Mali. She misses her mother. Yet she is determined to fight. Her story is going to be a key part of the film. It has to be. It is a story of pain and resistance. A warrior woman!

I want to capture on film the sense of sisterhood between Alice and Aminata that I witnessed on their first meeting. To this end, I decided to film the two of them strolling along the Thames. Alice and Aminata walked arm in arm, like old friends exchanging confidences. Alice's warm ways are appealing, and she made Aminata feel at ease quickly. It was the perfect way to begin Alice's journey within the film.

Since I have never had much film training, I use the opportunity of each successive film to try out at least one technique that is new to me as a way of learning about it. I decided that this time around I wanted to use a steadicam operator for the shoot on the river. A steadicam allows the camera to move in a way that a body can't move: into corners, around curves, and up and down. It does this in a smooth fashion that other types of handheld camerawork don't permit. Every shake of the hand in handheld camerawork is magnified in the projection of the film, and a steadicam can lessen this shake.

*12 November 1992*

Dear Alice,

These last ten days since I returned from California have been hectic. Everything is rapidly moving ahead on the film. I will fill you in as much as I can.

1. Efua is now our official consultant and will be paid a fee for this. Her help is invaluable in terms of contacts in Africa and elsewhere. It seems that her contacts are much better in The Gambia and Senegal. I think we should go to countries where our contacts are good, where it is easier and safer to film. We also want to go where there are powerful women campaigners against female genital mutilation.

The Gambia: Efua knows a woman who is like a sister to her and who has been working against female genital mutilation for many years. She will be able to direct us to grass roots contacts and also talk to us about the overall situation. Efua said she would welcome us, ease official stuff, and generally look after us. Exactly what we need. The work that the local organization is doing could be a positive model for other African countries trying to effect change.

Senegal is the spirit of Africa, Efua says, and we have to go there! Awa Thiam is based in Senegal and, as you know, is a militant campaigner against female genital mutilation. But apparently she is suspicious of outsiders and has refused to give interviews. It would be great if we could get her agreement. I am just reading her book again, and I think she would offer interesting input. Her perspective is radical, and she links female genital mutilation to patriarchal control and violence against women and doesn't see it as just a health issue, as some of the other campaigners do.

You probably know all this anyway. But for me, so much of this information is new. I am doing a lot of research and reading at the moment, getting angry and feeling pain at the pervasiveness of this violence against so many women and girl children. At the same time, I am excited about all the possibilities for the film.

Efua has also given us a contact for Dr. Henriette Kouyate, a gynecologist who is working against female genital mutilation. Dr. Kouyate runs a clinic in Dakar, where we could film her with some of the women in her practice who have been mutilated.

Youssou N'Dour, that wonderful musician, also lives in Senegal, and a friend, Isaac Julien, knows him and has offered to call him for us if we want.

I suggest we write a joint letter to Awa, asking if she will cooperate with us for this film. There is an urgency to this, because we need to finalize which countries we are going to, so that we can start applying for the official permissions from the relevant governments, which can take months to process.

2. There are two women in the U.S. whom we may also be interested in talking to and/or interviewing:

Assitan Diallo: An academic from Mali based in the U.S., she has done considerable research for a thesis on genital mutilation in Mali among the Dogons. Efua thinks we should talk to her, as she may be helpful if we go to Mali. We are trying to find a contact number for her but haven't been successful so far.

Seble Dawit: She is currently writing a book on female genital mutilation from a human rights perspective and has toured Senegal, Kenya, Egypt, Ethiopia, and Uganda to research it. She is a consultant in human rights to a variety of organizations and lives in New York.

Finally, I have been interviewing for a researcher/production assistant to accompany me to Africa. So far I have not come across anyone I think is appropriate: they are too young and/or they see it as a mere job. I would like to have someone who is also committed to the subject matter, or at least sympathetic to what we are doing. I am also concerned about my safety and want to have someone who will not be easily pushed around! Now, I wondered whether Deborah would fit this role. I think she would definitely be able to deal with officials in Africa without being bullied, and what she doesn't know about film production in terms of technical know-how she would make up for through her commitment and understanding of the subject matter. This is a wild idea, and I just want to air it with you and see what you think. This person would accompany me to Africa, and we would be there three weeks, setting up everything, meeting all the potential interviewees, looking at locations, and clearing all official channels. She would then stay on for an additional two weeks when you and the crew arrive for the filming. So all in all, it would be five weeks, for which, of course, she would get paid. If not Deborah, is there someone else you think might be good? Again, since the plan is to leave for Africa on 18 January, we need to find this person very soon.

I have come up with a revised production schedule, because it seems that February and March are better months as far as the weather is concerned. Later on, it can get too hot and also wet. Let me know what you think about this change.

### Production Schedule for **Warrior Marks:** *Africa*

*18 January 1993:* Pratibha Parmar and production manager to Senegal and The Gambia for a three-week research and setting-up period. The time will be split equally between the two countries.

*5 February 1993:* Alice Walker arrives in London en route to Africa.

*8 February 1993:* Alice Walker and full crew leave together for The Gambia. One day's travel, one day's rest, and five filming days.

*9 February 1993:* Happy Birthday, Alice.

*11 February 1993:* Happy Birthday, Pratibha. (I like the idea of having our birthdays in The Gambia.)

*15 February 1993:* Pratibha, Alice, and full crew to Senegal. One day's travel, one day's rest, and five filming days.

*22 February 1993:* Everyone returns from Senegal to London.

*25 February 1993:* Alice returns to California.

Alice, please note that you can come earlier to Africa and leave later if you wish to do so. We can finalize the dates in December so that the bookings for the tickets can be done in advance and all the appropriate inoculations completed in time. (I am not looking forward to these!)

Well, Alice, there are a lot of things to think about, talk about, and make decisions about when I come to California.

I will be forwarding a detailed shooting schedule for 1 and 2 December in northern California very shortly. Suffice it to say, the crew is booked and so are the overnight accommodations. I will be coming up on Sunday, 29 November, in order to work out where and what I need to film before the crew arrives. Looking forward to a California hot tub!

I will call as soon as I get into town. Meanwhile I send you lots of

Love,<br>
*Pratibha*

NOVEMBER–DECEMBER 1992.
MONTREAL/SAN FRANCISCO/LOS ANGELES

I flew to Montreal, where the Montreal Lesbian and Gay Film Festival was showing a double bill of *A Place of Rage* and *Double the Trouble, Twice the Fun.* The latter video explores the ways in which lesbians and gay men with disabilities experience the tyranny of the body beautiful prevalent within gay culture. I had completed that video in June 1992 for the Channel 4 series called *Out.* The best compliment I had received about it was from Derek Jarman, one of my favorite filmmakers, who phoned me after the television broadcast, telling me how much he was moved by it. He said it was beautiful and painterly. This was praise indeed from someone I respected and admired a great deal.

Canada was cold, in fact freezing, but I thoroughly enjoyed my brief trip. The response to my films was energizing. I realized how important these festivals are to independent filmmakers. This is where we get our nurturing and affirmation as well as criticism. I like being asked questions after screenings. They say so much about what works in the film and what doesn't, and always serve to sharpen my thinking about the next film. Derek Jarman has said that "anyone who works with any degree of commitment will find they want an active audience."

From Montreal I flew to the warmth of San Francisco to begin a two-day shoot in northern California, where Alice lives. This shoot was a real pleasure. I worked mostly with the same crew as on *A Place of Rage,* and there was an easygoing camaraderie among us.

As I looked down into the valleys from Alice's house,

the mist rising onto the cedar trees and the redwoods, the magic of this part of California captivated me. It looked like a Japanese painting, delicate and poetic. One of my favorite places in the world! I would retire here if I could be assured that I could get good Indian food!

I want to establish the start of Alice's journey and interview her in her own setting. What made her want to write *Possessing the Secret of Joy?* When did she first hear about female genital mutilation? What kind of responses has she had to the book? What are her expectations of Africa?

A few weeks before leaving England for California, Alice sent me a script that she had written for our film. She called it *Like the Pupil of an Eye: Genital Mutilation and the Sexual Blinding of Women.*

Reading the script made me realize yet again how vulnerable Alice makes herself to the world through her words. The simple elegance of her writing, and the painful honesty with which she confronts having been blinded in one eye by her brother, is a complex and powerful combination.

During the California shoot, Alice spoke with wisdom about taboo subjects and deeply difficult personal issues. We filmed a few takes of Alice by the fireside, reading and telling her story. We also filmed her with her dog, Mbele. Alice was patient while we changed film magazines, which last only ten minutes. We spent too much time trying to get the lighting right!

The direct telling of her story to the camera was engaging. I knew, as the camera was rolling and Alice was speaking, that this telling will be another layer driving the narrative of the film forward. I like creating layers upon layers, narratives within narratives, trusting that these will inform,

*Pratibha directing Alice in California.*

|||||||||||||

*Cinematographer Nancy Schiesari (right) with assistant camera operator Brenda Haggart.*

|||||||||||||

evoke, complement, and come together in dynamic and forceful ways: images and words that shimmer with meanings, like a poem.

With each step I found that this film was growing within me in a visceral and intuitive way. The filming and reading I'd been doing about this subject were taking root. Lacking formal training in film, I have evolved my own method and personal sense of aesthetic strategy. I try to find within each film its own poetic meaning, its own "poetic logic," as the Russian filmmaker Andrei Tarkovsky said. For me, filmmaking is like collecting different treasured elements and fragments that eventually form part of a whole; it is not prescripted. At the time of filming, I might not necessarily know how a particular element is going to fit into the whole. I often like to go with my instincts, filming images and words that will be given form in the editing period. It's this aspect of making documentary films that I love so much.

The next day, after doing a few pickup shots of Alice on the land, we returned to San Francisco, but not before getting our car towed out of the mud. I drove back to the city with Kerin and Sharin Sandhu, twin sisters who have been helping out on the shoot. We listened to old Indian film songs and attempted to sing along in our broken Hindi as the torrential rains came down.

Although exhausted from my traveling and filming, I had to go on to Los Angeles for the screening of *A Place of Rage*. Nazila Hedayat, the codirector of the Los Angeles Lesbian and Gay Film Festival, had organized this event with women from the local black lesbian group.

Unfortunately, a houseful of three hundred women were as disappointed as I was when the projector broke down.

Los Angeles is really not geared up for 16mm films! Film-makers must go through such nightmares at least once in their lifetime, I was told, but it didn't make me feel any better.

It's curious how one mishap can lead to a fortunate encounter. I talked to Nazila about looking for a French-speaking production manager to take with me to Africa. I had little time, and I had been interviewing and talking to a number of women. Nazila offered herself. Immediately the idea was appealing, because I could easily imagine spending a month with her.

I decided to think about it, as I had to have someone with production experience as well as a commitment to the film. Shooting in Africa would not be easy, and at the end of the day the person who accompanied me would have to have tenacity and determination. I needed a right-hand person who would be my friend as well as a professional colleague. Nazila was obviously an efficient and hardworking person in her capacity as codirector of the film festival, but I didn't really know her personally.

A week later, she called me in San Francisco, and I decided to take her on as production manager for the Africa shoot. Nazila had read Alice's book and was then reading Awa Thiam's. She seemed serious and eager, and she said she wanted this opportunity to work with me. Above all she was willing to take the risk of not going back to her current job and was giving up her apartment. What more could I have asked by way of commitment—and she was an Aquarian!

Again I felt lucky and relieved. This project had so far attracted some goodhearted and committed people.

## 1 January – 23 January 1993. London

The three weeks of preproduction in London before leaving for Africa were relentlessly hard work. Nazila was wonderful to be with; working with her was a joy.

Alice and I had been trying for several months to contact Awa Thiam, as I'd wanted to talk to her before we left for Senegal, but we'd had no luck with this. Finally, just before we left London, Nazila and I reached her by phone and for the first time we spoke: I in my broken French and Awa, also in French, responding slowly for my benefit. We fixed up a telephone date with her for the next day.

We found her to be warm and welcoming, not at all what we had been led to expect. She even wanted to meet us at the airport. She would be our key and crucial contact in Senegal. It was a miracle that we had found her after months of searching and phoning, and just as we are about to leave. What an elusive person she seems!

Nazila and I left for Senegal on Saturday, 23 January, with Alice and the crew scheduled to arrive two weeks later. We were going ahead of the crew to set everything up and do the necessary research on location. I was hoping to have three weeks in Africa for preparation and research, but for financial reasons, it was not possible to do this. Still, we had two weeks. This time was essential, since there was only so much that could be done on the phone and through letters.

Before we left, I had received some great news: I'd been selected to receive the Frameline Award at San Francisco's International Lesbian and Gay Film Festival. It's an award presented to an individual who has made "an outstanding contribution to lesbian and gay media." This was wonderful news. It was reassuring to know that the difficult and complex issues I try to address in my films are being affirmed and acknowledged. What better news than this could I have had to carry me through Africa and help me face the coming month with renewed energy and confidence.

I looked forward to being in Africa again, to touching the ground of the continent I was born in, to meeting Awa Thiam, as well as other women whose anger and pain were moving them to fight the deep-seated traditions so harmful to young girls. And I contemplated how emotionally difficult this journey would be, but I promised myself I would do all I could to make a film that was tender yet angry, informative yet emotional, a film that would contribute to change, however limited.

*. . . who would believe*
*the fantastic and terrible story of all of our survival*
*those who were never meant*
*to survive.*

*Joy Harjo*

SATURDAY, 23 JANUARY 1993

Nazila and I arrived in Dakar exhausted after twelve hours on the plane. It was supposed to have been a direct flight from Brussels, but the plane stopped in Conakry, Guinea, for two hours. It was here that the transition from the dominant whiteness of Europe to the Africa of Africans happened. Most of the white people got off the plane, and a large group of African men boarded the flight for Dakar.

It's been six years since my last visit to Africa, and that was to Kenya, twenty years after my family left Nairobi for England, in 1967. I'd spent the first eleven years of my life in Kenya, and Africa remains in my body's memory.

The smells, the colors, the light, the red-earth roads, the people, all evoked childhood memories of Mombasa, the port town in Kenya, where my grandparents settled after leaving India in the 1940s. The Indian migration to Africa and parts of the Caribbean, all former British "colonies," was extensive. The British "facilitated" this migration so that Indians could work as indentured laborers building railways, or on plantations owned by the British "sahibs."

On arrival in Dakar, we managed to get through customs and immigration without too much hassle, except that Nazila got stopped and questioned. She had warned me that this often happens to her. Immigration officers around the world are not used to seeing UN refugee passports, and Nazila is an Iranian refugee. She thinks the officials stop her out of curiosity more than anything else.

The taxi that was supposed to take us from the airport to the hotel had to be push-started. Its driver had a lined face that looked like it had seen much hardship. He was a sweet man who at sixty-two was the sole wage earner for his large family. He complained about the severe unemployment in Senegal, especially among men, and the general level of poverty.

Everywhere there were billboards with photographs of President Abdou Diouf, signs of impending elections. I wondered how this would affect our filming plans.

The hotel is grubby, yet expensive, and the men at the reception desk have been quite rude to us . . . I don't think they like women unaccompanied by men. The taxi driver certainly was right when he said that this hotel is overrated!

Nazila and I are sharing a room (separate rooms are too expensive). Traveling with her has been fun. We are getting to know each other, exchanging life stories. We are making a connection as exiles who have lived in many countries, and we've discovered that we are both nomads at heart.

### SUNDAY, 24 JANUARY 1993

Slept like a babe last night but had a strange dream! In the dream, I woke up and looked out the window. The sky was black; it was raining and desolate. I cried out that this was not Africa and there was no way I could make a film here. I wonder if this is an anxiety dream or an omen of some sort.

Today was a hot, hazy day, and the reality of being in Africa after planning the trip for over six months finally hit me as I looked out the hotel window and saw men in long flowing robes sitting around the taxi stand at the entrance to the hotel and women walking by with pots confidently balanced on their heads.

I am about to embark on the most important part of the shoot, and I don't understand why I am feeling so calm and confident, as if everything is going to be fine and happen just as it should.

After breakfasting on coffee and croissants, we scouted the town and all the hotels in Dakar. Tomorrow we will move to a cheaper and friendlier place. Everywhere we go, we are stared at. We were also followed, by a group of men who wanted money for the junk jewelry they were selling. We managed to get away before things turned nasty.

A few women were selling fruit at a market stall, looking regal and proud. Women flower sellers walked around the market square with brightly colored flowers of gold, purple, and red displayed on their heads in wonderful arrangements. Much more inventive than anything even Fellini could have imagined.

Sitting in a doorway was a woman with all her belongings in a few bags and a suitcase. She sat there with the sun on her face, writing in a book with great concentration. She looked up, saw me watching her, and we both smiled in recognition.

It's difficult to grasp that so many of these women could well be genitally mutilated. According to the only available statistics, 50 percent of all women in Senegal are.*

*Possible sequence in the film:*
A series of sound bites from different women, each of them reciting a common justification for the continuation of female genital mutilation.

*Woman 1:* "They say it's in the religion, but there is no mention of excision in the Koran."

*Woman 2:* "They say the clitoris is dirty and has to be destroyed."

*Woman 3:* "They say the clitoris would grow like a penis and hang between the legs if it is not removed."

---

*From The Hosken Report: Genital and Sexual Mutilation of Females* by Fran Hosken.

*Woman 4:* "They say that a woman will remain childless if she is not excised."

*Woman 5:* "The clitoris is an evil, which makes men impotent and kills children at birth."


## Monday, 25 January 1993

What a day! Changed hotels early this morning and finally made contact with Awa Thiam. She had apparently come to the hotel yesterday with a friend, looking for us, but those jerks in reception couldn't find our names on the register.

It took more than an hour to change money. Reminds me of the time in India when it took an hour to buy a stamp and post a letter!

Awa's office is in an extension of the Institut Fundamental d'Afrique. Awa is very, very tall and quite imposing. We spent three hours with her, explaining the intentions of the film, who Alice is, and why she is involved. For the first two hours, I was convinced that she was not going to cooperate. She was difficult to read and seemed suspicious of us. I understand why she is reluctant, since we have only just met, but why was she so friendly on the phone before we left England? Why did she agree to be filmed and to introduce us to the autonomous women's group she is involved in?

In the end, Awa was gracious, yet distant, unwilling to give us any concrete commitment. She seemed surprised to learn my age and that I am a filmmaker. She thought I was about twenty years old, which I suppose should be flattering, since I am almost twice that, but is somehow annoying.

Eventually we managed to find some common ground, and she began to talk more openly. She was critical of the "softly, softly" approach of the Inter-African Committee. She feels they work against the autonomous women's groups who are fighting female genital mutilation. The Inter-African Committee on Traditional Practices Affecting the Health of Women and Children (IAC) was formed in 1984 by delegates to the Conference on Traditional Practices in Dakar. The IAC, with headquarters in Geneva, has developed national chapters or groups in many African countries, including Senegal.

Awa has decided that working exclusively with women's organizations is no longer sufficient. She has joined one of the opposition political parties, Partie Democratique Senegalais (PDS), and is working with male politicians in the hope that the changes she is striving for in the lives of Senegalese women will happen sooner and on a larger scale. Her analysis is that power resides with men and in particular with the religious leaders. She says that if tomorrow all the religious leaders in Senegal decreed that female genital mutilation should be abolished, it would disappear within a few years. Clearly she is hoping to influence these religious leaders, many of whom are also political spokesmen. But still she remains a member of the Commission for the Abolition of Sexual Mutilation, an autonomous women's group she founded in 1982. These women fight against all forms of mutilation and publish the magazine *Women and Society*, which is probably the only feminist magazine in West Africa.

As a token of my respect for her writings and her work, I gave Awa a beautiful blue Indian shawl I'd bought in London. I also gave Awa a video copy of *A Place of Rage.* She kindly autographed my copy of her book, *Black Sisters, Speak Out: Feminism and Oppression in Black Africa.* After publication, there had been an enormous backlash from men and women in Africa, who castigated Awa for her radicalism and for speaking out so candidly in the book about female genital mutilation, polygamy, child marriages, and other forms of patriarchal oppression specific to Africa. *Black Sisters, Speak Out* inspired me to believe that as a woman and a feminist, I have a moral and political duty to be concerned about genital mutilation. Her relentless anger and her logical arguments demand from the reader a commitment to fight for fundamental change and equality for women. Awa is one of the most important reasons we had for coming to Senegal.

The big question now is will she be able to help us? She says she will let us know tomorrow morning, after she has been informed of her schedule for preelection lobbying in the rural areas for the PDS.

After our meeting with Awa, Nazila and I went for a walk in the city heat. The guidebooks say that Dakar is supposed to be the Paris of Africa, yet we saw none of the wealth and opulence of Paris. What we did see were little French patisseries and ice cream parlors, which expatriate French families frequent in the evenings.

The street life in Dakar is reminiscent of India, with people hanging out on corners and cruising slowly down the wide streets as the sun goes down and the city cools off. We went for coffee in a restaurant and suddenly realized why many of the women were eyeing us with suspicion: they were young African prostitutes, and suspected us of horning in on their territory. Their clients are white men, tourists who come to Africa specifically to have sex with African women. It is so sad that these women must resort to prostitution as a means of survival.

While we sat drinking our coffee, Nazila asked me lots of questions about the film, about Alice's role in the film, about female genital mutilation, about my approach. Her intense questioning, while a bit irritating, forced me to focus on some unresolved issues surrounding the film.

I think Nazila had a hard day today! Still acclimatizing. I guess this is a big change from her life in Los Angeles!

In the evening, a sweet, gentle man played the *kora* and sang to us while we ate fish stew in a local restaurant recommended by the hotel porter.

Spoke with Shaheen on the phone. I miss her very much. I feel I have been away for centuries. Is the film any more real today? I really don't know. Nervous now but must meditate before going to sleep. At least here Nazila and I have separate rooms.


TUESDAY, 26 JANUARY 1993

The island of Gorée is twenty minutes off the coast of Senegal. You can only get there by ferry from Dakar. It is a historic place where hundreds of thousands of African women, children, and men enslaved along the west coast of Africa were kept before being shipped off to the Americas.

Today it is a very pretty place, with dusty pink and faded yellow houses draped in bougainvillea. No cars are allowed on the island, which you can traverse in a leisurely hour. We could see the beautiful coastline as we walked to the highest point, where we came across a small community of Rastafarians living in the makeshift shelters that have been there since the invasions by the Portuguese, then the Dutch, then the French, then the British. The list of colonizers is long and predictable.

Many of the families who live on the island have courtyards that open onto the road, and as we strolled around, we caught glimpses of women pounding grain and washing and dressing children. I noticed how proud the women were, how they held themselves with a noble majesty. These courtyard images reminded me of my childhood home in Nairobi.

The everyday life of the Gorée islanders is regularly interrupted by tourists, who arrive by the boatload twice a day. There is an uneasy acceptance of this by the locals, because tourism is their main source of income. On the ferry we met a graduate in engineering who is working as a tour guide because there is no other work available to him.

After a delicious fish lunch, we went to see the House of Slaves, our main reason for coming to the island.

In the December interview we filmed with Alice in California, she'd talked about the double tragedy of excised women who also faced enslavement. She wondered what had happened to pregnant women on board the slaveships with no one there to help them through childbirth. What pain did these excised and often infibulated women experience when they were raped by the slavemasters, something we know occurred with frequency? I had decided I would like to film Alice talking about this previously invisible aspect of the enslaved woman's painful experience. This house, a symbol of that history of enslavement, could not have been a more fitting location. I wanted to film Alice making a pilgrimage to Gorée Island, and in particular I wanted to shoot a final interview with her in the House of Slaves, the beginning of the journey for many African Americans. I wanted this sequence in the film to mark the end of our journey in Africa.

The House of Slaves is barbaric in its beauty. Built by the Dutch slavemasters, it stands dusty pink, framed by symmetrical, arched stairways. The ground level divides into small rooms, which once separated women, children, and men, and I imagined that the thick concrete walls had absorbed their screams of pain.

Here proud, dignified people were kept in chains before being shipped off, never to return. They were divided from their loved ones, many dying en route, others dying in the bonds of slavery.

There is a doorway leading to the ocean through which the water sparkles tantalizingly. It is from this point that the enslaved, chained and manacled, were pushed and shoved onto the ships.

Under the stairs is a space marked with a sign for "rebels," those who were punished even further for resisting, for refusing to listen to the masters' commands.

Upstairs, where the slavemasters lived, large windows look out to the ocean.

As I walked around the house, it was disturbing and painful to imagine what transpired there not so long ago. I shivered at what this setting elicited in my imagination: images of children, women, and men being herded into these

*The door through which
our ancestors passed.
Gorée Island, Senegal.*

||||||||||||||

*The history and
significance of
Gorée Island.*

||||||||||||||

### "BEHIND CLOSED DOORS"
### HOUSE OF SLAVES

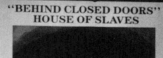

The Slave House (Maison des Esclaves)
was built by the Dutch in 1776. It is the
only remaining slave house on Gorée
Island. Through these doors are a number
of cells that held children, adults, and
"rebellious" slaves. They were usually
held here for a few months before
departure to some unknown land.

Gorée Island, West Africa

Researched and photographed by Beverly A. Davis
1990

animal pens. Nazila asked me: "Can you hear them screaming?"

That evening we drove to the outskirts of Dakar, where
Awa had invited us to a meeting of one of the opposition
parties, the PDS. All the women sat on the ground, while
the men sat on benches. The meeting took place in a courtyard, and as the sun went down and a crescent moon slowly
emerged, candles were lit so the secretary could read his
speech.

We were introduced to everyone at the meeting and were
greeted warmly. Of course, no mention was made of what
we are doing here and what our film is about. Awa says we
have to tread carefully. Later we were taken to meet the
special adviser to the PDS leader. He is Princeton-educated
and sounds it—all waffle and evasiveness, no concrete commitment.

The women in the PDS requested permission to talk to
the leader and ask him to speak out about women's issues
in his election campaign. The adviser was very patronizing
and pronounced tired old sentiments: Let's discuss the so-
cialist agenda, the larger picture, and later we will talk about
specifics, etc. Nazila got involved and asked some touchy
questions, but the leader remained cool.

I felt tired and wanted to get back to my grubby hotel
room. Awa made her excuses and left us, but not before
briefing us about tomorrow's trip to the village of Kavil. I
had indicated to Awa that I wanted to film in a village,
since the majority of excisions take place in rural areas, and
so she organized this trip for us.

Awa has done some work with women in villages, particularly in Kavil. She has arranged for us to meet the village midwife, who used to do circumcisions but has stopped, mainly due to Awa's intervention.

A PDS man took us to meet his daughter Adelaide, who will be our interpreter tomorrow in Kavil. He also wanted us to meet his first wife. (He has two—polygamy is rife even among the so-called enlightened men.) She turns out to be a very sweet woman, who invites us along with her other guest, Madame Fall, to lunch next week. Madame Fall, the organizer for the PDS women's section, has eleven children and has taken care of them all on her own. She dresses in beautiful batik gowns and talks a lot about sex. She is keen for us to meet her "leader."

Everyone has been welcoming, even when they know we are making a film about female genital mutilation. I had been prepared for more clandestine discussions, but it's not been necessary. This is a hopeful sign.

The last thing I needed this evening as I rushed back to the hotel to try to catch up on my sleep was to be stopped by the police. It seems I was driving the wrong way up a one-way street, but why this resulted in a one-hour interrogation I'll never know!


WEDNESDAY, 27 JANUARY 1993

I drove to the Kaolack region, to the village of Kavil. Nazila navigated and Adelaide sat in the back, too shy to talk. The roads were dusty, and the infamous Senegalese wind blew the dust everywhere. I reminded myself to talk to Nancy, the cinematographer, about the film stock she should bring with her. Filming in this climate is going to be challenging; the blazing sun bleaches out so much and creates many shadows, and the wind blows the sand and dust up in the air, making the landscape intensely hazy.

We stopped off at the house of Awa's sister, who is bourgeois, beautiful, and extremely sociable. Her husband is the owner of a local foundry, whose workers are all on strike. She had lunch made for us, and we all sat around the table, eating out of the same bowl. The dish was rice with a fish sauce. I have grown to like fish. I guess it's either that or starve.

The most interesting part of this visit was talking and videotaping Awa's fourteen-year-old niece and her school friend, who are acutely aware of women's rights issues and the differences between their comfortable lives and the lives of the majority of girls their age in the villages.

After the interviews, we were obliged to hang around for a few hours and be social. I got impatient. Finally we left with Adelaide, our interpreter, and her friend, whom we had been waiting for.

We arrived in Kavil and went to the compound Awa had described to us. It was comprised of a two-room home, an outhouse, and ground space for the goats and chickens to wander around in. We were greeted by the whole family; the mother (we never learned her name); her daughter, Hadija, and her two children, Hadija's brother, Jacques, and his friends from the village, and of course the dog and the goats. We gave the mother our letter of introduction from Awa, but she didn't read it.

After we parked within the fenced compound, Nazila spoke in French with Jacques, and we also communicated through our interpreter. Though somewhat bemused, the family was still welcoming.

Kavil is a small village situated on either side of a main

red-earth road, and every time a truck or bus goes by, clouds of red dust are whipped up and everything and everyone is covered by it. Children with ebony-black skins, parched by the sun, are layered with red dust. It's as arid as a desert, and the poverty is startling.

Nineteen-year-old Hadija does all the work. She has two daughters, aged three and five. We were shocked to discover that Hadija's mother is only thirty years old. She acts and looks like an old woman, and we attribute this to poverty. The daughter of the village chief, she is a witch doctor. She tells fortunes with cowry shells and then, for a price, suggests remedies.

We went to the well to help Hadija get water. There were only four taps, but more than a hundred young women and girls lined up with their buckets. The barren landscape and the absolute scarcity of basic necessities like water was disturbing. It reminded me of the time I spent in 1975 in a village in north India, working as a volunteer on a women's health development project. The poverty is the same, as is the harshness of the women's lives.

For the girls of the village, Nazila and I were of course the cause for much laughter and giggling. They must have wondered what we were doing there.

Later we sat outside in the compound, talking to Jacques, Hadija's brother, who seemed gentle and sensitive. When we asked him why he didn't help his sister by going to the well for water, since he is a big, strong boy and she has so many other things to do, he told us that he used to do this, but all the other boys in the village made fun of him, so he stopped.

The family wanted to hear music on their cassette player,

but there were no batteries. So we all piled into our hired car and went for an excursion to the nearest "town," whose name we couldn't ascertain. The one street in the town was full of prostitutes waiting for the passing truckdrivers, as well as any local men who came along. This trip to town was a big event for Hadija, who wore her best white skirt. She rarely gets a break from her daily workload. Nazila and I have really taken to her: she is incredibly resolute and determined.

As we drove back on the dark, dusty roads, we passed small buses full of people hanging on to the back doors and windows, the roof racks packed high with goods. Apparently the goods—basics like sugar, flour, clothes—have been smuggled in from The Gambia into Senegal. We have discovered that everything is much more expensive in Senegal.

I know this is one image that will stay with me for a long time: the star-filled African sky, the buses beautifully painted in bright colors full of people inside and out, traveling through the night.

Back in the village, the grandmother came out to the compound and put a tape into the cassette player, saying it was her favorite music. Incredibly, it was from an old Indian film I knew from my childhood, and I translated one of the songs for them. The intrusion of these familiar sounds into this otherwise strange setting was bizarre and incongruous, but in some ways it made sense.

Before going into the hut for the night, we'd noticed that the men were all sitting around the compound listening to music, while Hadija made tea for them. The men here seem to stay up late, get up late, and not do very much work!

That night before we went to sleep, something funny occurred: Awa had warned us not to drink the water or eat the food, so we'd eaten very lightly at the family meal, and were consequently starving. After ostensibly retiring for the evening, the four of us—Nazila, Adelaide, Adelaide's friend, and I—decided to sneak into our car and get some food. So there we were, gobbling down our half-eaten sandwiches from earlier in the day and taking turns drinking from the bottle of water we'd brought. We then snuck back in to our sleeping quarters. All that so as not to offend the family!

We had come armed with insect repellent to keep away the mosquitoes, and we sprayed it all around the mattress Hadija had laid for us, as well as on the exposed bits of our bodies. Nazila and I lay next to each other, giggling, trying not to feel claustrophobic, for Hadija had shut the door. Apparently, animals stray into the rooms at night if the corrugated-iron doors are not kept shut.

Nazila was telling me how a man we both knew from the film world had warned her that I was a diva and that she would find it difficult to work with me. She said she couldn't wait to tell him how this diva had spent the night in a shack on a flimsy mattress, and how many divas did he know who would do that? It's funny, but I've found that when you are direct and outspoken and make demands, you are labeled a diva. Men do it all the time, and they are just called men!

Suddenly there was a noise, and Nazila sat up and, with alarm in her voice, said, "Don't look, Pratibha—whatever you do, don't look." She sounded absolutely terrified. Of course I had to look. I put on my glasses and sat up, and there, not two feet away from us, where Adelaide had left the flashlight on, were cockroaches, hundreds of them, scuttling about. If we hadn't sprayed the edges of our mattress, they would have been on us in no time. It was like a scene from a B horror movie. We crept out of the hut like two thieves in the night and went into the car, this time for good.

I tried not to be afraid in the dark car, even though we were in the middle of Senegal and no one except Awa knew where we were. Every time a truck went by, we felt the dust rise up and creep in through the windows. While scary, it was also somehow magical to feel the millions of stars watching us as we waited for dawn.

At daybreak, Nazila and I walked to the bushes and used our last bit of water to brush our teeth. This proved to be delightfully refreshing.

Later, Nazila went through a purification ceremony. On the previous day, the grandmother of the family had told Nazila's fortune, saying she'd found evil spirits. These evil spirits needed to be exorcised, according to the grandmother, or Nazila would never get a man. Even though this was not a worry for Nazila, she decided to go through the purification anyway. (My fortune had warned that there was something in me that frightened men. Suits me fine to keep it that way.)

The purification ceremony consisted of all kinds of powders sprinkled into buckets of water and then poured onto Nazila, who crouched on the ground, completely naked. It's a good thing she is a nudist and didn't mind being undressed in front of all those people. I told her to keep all her orifices tightly shut. I certainly didn't want my production manager and right-hand person to get ill. I videotaped the whole thing on my Hi8 camera.

We had long discussions about female genital mutilation

with Hadija, her mother, and her brother, Jacques, and learned that in Kavil some families practiced it and some didn't. Hadija and her family didn't, and encouraged others not to. They confirmed what we had been hearing in Dakar—that it is practiced only by certain tribes.

Jacques was sent off to get the circumciser so that we could interview her. It turned out that she was still off in some other village, doing her work, so we arranged to return to Kavil on our journey back from The Gambia to Senegal. Jacques promised he would have the circumciser waiting for us, and would arrange for us to film an interview. I was relieved to leave but a bit sad to say goodbye to Hadija.

We drove back to Dakar. As soon as we got into the hotel, I jumped into a shower and scrubbed myself clean of all the red dust. I called room service for an omelet that I could guarantee would be greasy. I decided I would die for a curry!

Later we rushed off to a meeting of the PDS women's group Awa had invited us to. There were about sixty women there—incredibly strong, articulate, and fiery. They talked about their various ideas for self-employment: setting up deals with fishermen, buying and selling secondhand clothes, making furnaces. Economic survival was their priority. We also talked to them in general terms about the film, but Awa had warned us that we shouldn't mention excision too quickly, so we waited for her cue.

Finally one woman stood up and spoke out against excision. She said it was taboo to talk about sex, but that it was about time women like them began to talk about the consequences of genital mutilation. She herself had not been mutilated but had close women friends who have been excised, and have suffered a great deal because of it. Some of the older women called her dirty for speaking so openly about something they consider intimate and private. The difference in the generational response to this subject illustrates the change that is slowly happening.

I wish I'd had the crew with me. The meeting was packed, and the energy of the women was fantastic. However, we did get many of them to agree to be interviewed.

At last I feel that we are getting to have concrete and definite contacts. The film, while certainly not yet in final shape, is starting to live a little.

Awa still doesn't know her schedule for the preelection lobbying, which means she can't give us a commitment.


FRIDAY, 29 JANUARY 1993

Felt terribly groggy today. About to have my period. Nazila and I spent all morning on the phone. First, I rang Efua in London to tell her about our progress in Senegal. She said she was busy organizing a picket of Brent Council, to take place in a couple of days. The protest is against a motion by an African woman, a local government councillor, to ask the British government to legalize female genital mutilation as "a right specially for some African families who want to carry on the tradition while living in the U.K." A coalition of African women's organizations and their supporters are protesting against this outrageous motion and plan to picket the council on 1 February. This action by Councillor Poline Nyaga came about after BBC 1 televised, in January, a program on female genital mutilation, in which the well-known Nigerian writer, Buchi Emecheta, who lives in England, made some comments in defense of female genital mutilation.

In response, Alice sent the following statement from California:

> *As Matron of FORWARD, I write to express my strong and unequivocal support for the statement against Female Genital Mutilation placed before you by Efua Dorkenoo, Director of FORWARD International and Bisi Adeleye-Fayemi, Coordinator of Akina Mama wa Afrika.*
>
> *I am saddened beyond measure that an African woman has embraced the torture of African children.*
>
> *While grieving over Poline Nyaga's betrayal of the rights of the child, I remain resolved to do everything in my power to support the efforts of African women all over the world who say: If the child cries, something is wrong! No to Mutilation! No to betrayal! No to torture! Yes to happy, healthy, smiling, free children.*

Later I phoned Debra Hauer in London and asked her to organize a crew to film the protest and get some interviews with women on the picket line. This truly is an issue spreading out into the African diaspora.

(When I looked at the rushes, I was pleased with the interviews that Sophie Gardiner, the production manager in London, had conducted on the picket line with Comfort I. Ottah and with Bisi Adeleye-Fayemi, the coordinator of Akina Mama wa Afrika in London. Comfort, a practicing midwife from Nigeria, spoke about her disappointment at being betrayed by another African woman, Poline Nyaga. "Torture is not Culture," she shouted. Bisi spoke passionately about how female genital mutilation is at the root of the oppression of African women.)

Meanwhile, back in Dakar, we called Awa as arranged. She was still unable to give us a commitment, but she did go so far as to talk about money—how much were we willing to give the PDS women's group? We arranged to meet her at her office after our lunch with Madame Fall and her friends.

Spent an hour at the bank, trying to change money, which made us late for our lunch appointment.

Our hosts had gone to great trouble and were dressed beautifully. I felt shabby in my T-shirt and shalwars, the Indian trousers Shaheen had given me for this trip.

Nazila's lips went numb from the chili sauce. (Later she said she thought she had the runs. Oh, no. . . .)

For two hours after lunch, we had the most stimulating conversation thus far on the trip.

Madame Fall (for some reason everyone addresses her as "Madame") is fifty-four years of age, full of joie de vivre; she loves to talk. She spoke about genital mutilation and related numerous stories and anecdotes. It was one of those times I wished I'd had the camera, but on the other hand, Madame Fall wouldn't have been as relaxed had the crew been with me.

She told us a story about one of the past presidents of Mali, who on a state visit to Senegal asked for his food to be cooked by excised women only. To be certain, he sent for women from Mali, who he *knew* had been excised. According to him, women who were not excised were unclean and therefore not fit to cook food for him.

I asked Madame Fall to elaborate on her comment that

increasingly men in Senegal were not interested in marrying women who had been mutilated. She said that particularly in the cities, more men are willing to come out and say they enjoy sex with women who have not been excised because this makes mutual enjoyment and pleasure possible.

This reminded me of an interview I'd read in Hanny Lightfoot-Klein's book *Prisoners of Ritual,* in which a man talks about how unhappy he and his wife are when they try to make love because of the pain she suffers as the result of infibulation.

The book says that some men go to the midwife and have their wives opened up, while others do it themselves using razors or knives. But many of the men say their sex lives are generally difficult:

> *Most men, when they talk among themselves, try to create the impression that they were able to penetrate their wives in the first night. . . . I know for a fact that this is virtually impossible. You are dealing with heavy scar tissue that is over-grown, and you are using flesh to penetrate it, and not iron. You could not penetrate a wall with flesh, and this is like a wall. Actually most men are afraid. . . .*
>
> *A friend of mine got abrasions all over his penis—for being such a fool. Most of the men who try to get through the first day get very drunk so they will not feel what they are doing, because they know they are doing something wrong. When they are drunk, they do not care what their partners are suffering.*

Madame Fall has agreed to arrange interviews for us, and all dates, times, and payments have been sorted out. What a relief! It's been difficult getting to this point in our journey, but our efforts are bearing fruit.

I am surprised and delighted by the openness of many of the women we have met, especially Madame Fall. When I told her why Alice and I are interested in making a film about female genital mutilation, she was supportive and encouraging. She said she'd enjoyed meeting Nazila and me and likes to see women making their way in the world. She is certainly doing that!

After our extended lunch, we went to see Awa. She was at first noncommittal, but then she said she wanted us to organize a press conference with the leader of her party, the PDS, and invite Alice to talk about female genital mutilation there. That is what *she* would like us to film. When we try to get concrete details about venues and dates for this press conference, it emerges that the leader may not be in Dakar when we are scheduled to film here.

I told Awa that I would rather film the autonomous women's group that constitutes the Commission for the Abolition of Sexual Mutilation than film a political leader who is not interested in doing anything about women's issues. And yes, while the overall framework of the film involves the oppression of women, the specific focus is on female genital mutilation as a manifestation of patriarchal oppression.

We went round and round and round, and I left with my head spinning. Will we ever get an interview with her, or is she going to disappear into the countryside, onto the campaign trail of this "leader" who refuses to speak out about women's oppression? She is the one person with whom I most wanted an interview, and one of the main reasons for our trip to Senegal. It's crucial to have voices of African feminists and activists like Awa who have been absent from previous documentaries on this subject.

The severe backlash and isolation that Awa experienced when her book was published have made her understandably hesitant to again speak publicly about female genital mutilation. Perhaps it is part of her new political strategy to talk about it in conciliatory and academic terms rather than in the angry, confrontational tone of her book.

While I tried to deal with my frustration at Awa's noncommittal behavior, I reminded myself of Efua's comment during Alice's interview with her in London: "Women can be killed" for talking about this issue. Perhaps Awa is being careful or perhaps she is behaving like a politician. I can understand all these reasons, but it would alleviate a great deal of stress if she would give us an answer, one way or the other.

Tomorrow we leave for The Gambia, and while Awa is not confirmed, Madame Fall has agreed to help us by arranging for the PDS women's group to be filmed and to find a circumciser who is willing to be interviewed on film.

Nazila and I ended the day with a budget meeting.


SATURDAY, 30 JANUARY 1993

We flew to Banjul, capital of The Gambia. We felt relieved to leave Dakar, which has been tiring and difficult. We were met at the airport by a sweet young woman who was sent by our local contact person, Bilaela. Since we had yet to meet Bilaela in person, we stopped off to introduce ourselves on our way to the hotel. One of the first things she asked was when she would get paid for her work in setting up interviews and visits to the villages for us. Right away, I sensed that something didn't feel right about Bilaela.

Nazila and I are sharing a room at an expensive hotel, the only one in The Gambia with telephones in the rooms, but at least Bilaela has booked us in here at a reduced rate.


## SUNDAY, 31 JANUARY 1993

Our first real meeting with Bilaela went badly. I felt cheated and upset that she had not been straightforward. We'd made a monetary agreement by phone from London, which I'd confirmed in writing by fax. This had outlined very clearly how much money we had set aside for Bilaela to administer for our expenses in The Gambia and what it would be used for. Bilaela now says she has spent all the money we had allocated. This means that we will be forced to cut down our shoot days from five to four, since we don't have spare cash tucked up our sleeves, as Bilaela seems to think.

I am afraid Bilaela thinks we are like the crew members from a major U.S. television network who were here three weeks ago. According to Bilaela, they had plenty of money, stayed in five-star hotels, and had many rest days by the pool and on the beach.

Bilaela was the American crew's local liaison, and apparently had problems getting the money they owed her and thinks we are probably not going to pay her either. I explained that we are not a big corporation, like the one from New York, but a small independent production company. I told her about Alice and how some of the proceeds from the sales of *Possessing the Secret of Joy* are going into the film.

Bilaela said she had saved some of the best "things" for us and that if we were willing to pay for some girls to be excised, she could arrange for us to film it. She said she'd done that for the New York crew. I couldn't believe what I was hearing. Female genital mutilation has become a media commodity! Her justification for doing this was that these girls would be excised anyway when their parents had saved enough money, so why not do it now when it could provide such "good" footage.

She went on to describe the scenario as it had occurred with the U.S. film crew: The circumciser had had two rooms, separated by a curtain. The girls were taken in, one by one, which was filmed. They'd recorded the screaming and then had filmed the girls being taken out again.

I told Bilaela that I was not interested in colluding in the torture and mutilation of girls for the sake of "good" footage. There was no way I would film this. There are many ways to show the horror of excision, but this was most definitely not the way I wanted to represent it.

It is easy for filmmakers to become part of the problem even while trying to resolve it. Even so, I was horrified to learn about what the New York crew had done. There seemed to be no rational excuse or acceptable justification for it.

Fatima, the interpreter Bilaela has arranged to be with us for the duration of the shoot, also talked about the New York crew. She'd been their interpreter too. She told us how lucky the New York crew had been because they'd come at "the season for mutilations." How could anyone consider this lucky? What some film crews would do to get "footage," and how distant and detached they must be from their subject matter is incomprehensible to me.

In the early evening, we went with Fatima to the village of Dar Salamay, which we'd been told was some sixty kilometers from Banjul. On the way we stopped at a market, in the town of Serekunda, to buy a bag of rice and kola nuts as gifts for the villagers, as Bilaela had advised. The road was bumpy and dusty. It was difficult to concentrate because I felt demoralized by Bilaela and devastated by what she'd told me.

On arrival in the village, Fatima took us to a courtyard to await the women elders. As I stood there, I decided I'd like to film in this courtyard, which is a hive of activity. It seems a veritable world of women was contained in that small space: women pounding and preparing food, women breast-feeding babies, little girls bringing buckets of water from the well, old grandmothers chewing bark and cleaning their teeth. This village was an excellent location from which to capture on film the daily tasks of women in rural Africa.

Fatima explained to the elders that we would like to return in a week with our film crew for the "coming out" ceremony that would take place here. The ceremony is traditionally held to celebrate the coming home of girls excised two weeks before. They are taken to a secret place on the outskirts of the village, where they are excised and taught about their role as future wives and mothers, as well as how to respect their elders. This rite of passage is marked by a ritual ceremony, during which all the women of the village dance and eat with the girls upon their return to their families. We had come here to negotiate the filming of this ceremony with the women elders, especially the circumciser. The circumciser at first stared at us intensely, then agreed. We gave them the bag of rice and the kola nuts, and a couple of the women broke into a dance.

We returned to the hotel after dark, hungry and tired. Nazila attempted to keep up my spirits, but I felt thoroughly depressed and angry. I had so hoped that The Gambia would be easier than Senegal.

Bad menstrual cramps don't help. I'd left England only a week ago, but it feels like I have been away months. Homesick for the first time and missing Shaheen. Went to bed hungry, since there was another fiasco with room service.

### Monday, 1 February 1993

Everywhere we go, we get hassled by men. It's driving Naz crazier than me. Bilaela was supposed to have sent a car to pick us up, but it didn't arrive. She is clearly as fed up with us as we are with her. I made several calls to Bilaela to find out where our car was, but she was not at home. She had just moved to new offices, and I didn't have the new address. Some detective work finally led us to her new office, and Bilaela was clearly shocked to see us.

I would have told her where to get off, but she is our sole contact here in The Gambia. We need her, and she knows that. In the office, Bilaela ordered everyone around. Her poor clerk wouldn't even open his mouth and speak without looking for permission from her first.

(In retrospect, Bilaela's story is very sad to me. I have tried to make a bridge of understanding between us. She is someone who has been courageous in taking up the campaign against female genital mutilation in The Gambia, and she still appears to be committed to it. But whether it is the emotional strain of working against difficult odds or the circumstances in which she is forced to work, it is upsetting to see how bitter and suspicious she is.

While we were working together in The Gambia, I had to keep reminding Bilaela to remember the spirit of the film. When I said that we hoped to make the film widely available and show it at film festivals, on television, in educational institutions, and before political organizations, she concluded that we were going to make a lot of money from it. I told her that Alice and I wanted the film to help raise funds for organizations working against female genital mutilation and we wouldn't be making money from it for ourselves. I told her that in order to change women's lives for the better, we needed to spread information about female mutilation everywhere we could. Bilaela listened and looked unbelievingly at me while I spoke. She didn't say anything.)

While we were in the office, a phone call came from New York. It was Stephanie, the producer-director of the American news segment on female genital mutilation. Bilaela had been very much involved in this film and yet she didn't tell Stephanie that we were sitting right there in the office with her. Stephanie had somehow found out that Bilaela was working with another film crew to make a film about female genital mutilation, and was asking Bilaela who we were and when our film would be shown. She was pressuring Bilaela not to cooperate with us. We heard Bilaela say that she had not given the American crew an exclusive and therefore felt free to work with us. She ended the conversation by saying, "Please don't cry, Stephanie. Of course I love you too, Stephanie, and don't worry about it."

I emphasized to Bilaela that as far as we were concerned, the more attention there is to this subject, the better. She can't understand that we do not feel competitive about it. The fact that I came into filmmaking not as a journalist or a current affairs programmer but as a political activist means that I have not got the same awful mind-set of so many media people, who get trapped into seeing everything as a story that someone else might get first.

Having spent all morning and half the afternoon with Bilaela, we decided to leave her office and have lunch in a café recommended by the guidebook. Fish again!

We walked around Banjul for the rest of the afternoon, thoroughly disgruntled and quite sad. It was hard to believe that Banjul is the capital city of The Gambia. There are few paved roads; consequently everything is dusty. Many people live on the streets. The stores are like shacks. Banjul seems so much poorer than Dakar.

When the British left The Gambia just before independence in 1965, they deliberately destroyed many of the buildings and roads as an expression of their anger at having to leave. In India, too, so many of the temple arts, sculptures, and historic buildings were destroyed by the British before they left India after being there for four hundred years. All over the world, similar signs of colonial devastation abound. Aime Cesaire summed it up in *Discourse on Colonialism:* "They talk to me of progress, of 'achievements,' diseases cured, improved standards of living. I am talking about societies drained of their essence, cultures trampled underfoot, institutions undermined, lands confiscated, religions smashed, magnificent artistic creations destroyed, extraordinary possibilities wiped out."

## TUESDAY, 2 FEBRUARY 1993

It was great to talk to Shaheen on the phone this morning. She was clear and supportive and encouraged me to find a way of making all these experiences part of the film. I have never made a film in this way before; the research on the ground is so dependent on other people's contacts and commitments, including people whose motives I am not always comfortable with.

I spoke to Alice, who had stopped off at my house in London on her way here to Africa. She sounded very excited about coming to Africa, and I wondered how she would respond to everything here. I mustn't bombard her with all my anxieties and cynicism when she arrives with the crew on Friday. She hasn't been to Africa since 1967, and then she was in East Africa. I want this trip to be memorable for her.

Received a letter from Wendy, a friend who lives in Montreal. Her words of encouragement and understanding lifted my spirits, but my body continued to ache all over. I feel weak from my period and have a cold sore on my lip. It must be a sign that I am run down. The stress is getting to me.

I started the day off with reflexology at the clinic in the hotel. It is run by two Scottish sisters who have lived in The Gambia for over ten years. Rose, one of the sisters, came here from England on holiday with her husband ten years ago and fell in love with Malign, their tour guide. She then left her husband and grown-up kids to be with Ma-lign. Clearly, she loves The Gambia and has made it her home. How romantic!

I felt so much better and less anxious after my reflexol-ogy treatment. Nazila and I phoned Awa, who still wouldn't give us a decision. We were trying to put together a shoot-ing schedule, and her indecision was not helpful. We also made contact with Dr. Kouyate in Dakar, to whom Efua had referred us. She is the general secretary of the Sene-galese National Committee on Traditional Practices, a branch of the Inter-African Committee on Traditional Practices. Dr. Kouyate sounded helpful and asked us to fax her a sum-mary of the film. I prepared one in English, which Nazila then translated into French. We don't have a typewriter, so Nazila transcribed it slowly in her best French handwriting. Ah, the joys of productions on location, with no office or secretarial backup! We rushed to get it to Dr. Kouyate, since time is of the essence now, with Alice and the crew arriving in two days' time. I wish we had contacted Dr. Kouyate before, but I was so hoping that Awa would come through. I still am!

We'd sat on our balcony, overlooking the gardens, as we wrote our letter to Dr. Kouyate. Beautiful and rare Af-rican birds came and sat near us. It is a source of joy to see these birds so free. People from all over the world come bird-watching in The Gambia.

It was good to have a precious moment of quiet for a change. And while Nazila and I have gotten used to being around each other, I think we both look forward to having our own rooms.

We finally got a fax from the Ministry of Culture and

Communication in Senegal. This is in response to the fax I sent from London before leaving for Africa, in which we requested official permission to film in Senegal. Most governments around the world have to vet foreign film crews before they will give them permission to enter. Since they never replied to us, we decided to come anyway, especially since Awa reassured us that we would be in Senegal at the invitation of her women's organization. Can you believe it? The Minister of Culture wants yet more details about who we are going to see and film in Senegal. This is so stressful. We can't be too upfront about the subject of the film, as the government is particularly sensitive to foreign crews making films that they feel will show Senegal in a negative light. Yet we need to tell them something, otherwise we may get stopped at the border between The Gambia and Senegal and not be allowed in. We can't tell them about Madame Fall and her group, because they are part of the opposition party, and I am sure we would not get permission if we mentioned them. What to do?

WEDNESDAY, 3 FEBRUARY 1993

Immediately after breakfast, Nazila and I negotiated with the manager of our hotel for a package deal including reduced room rates for when the crew arrives. This will help us to stay within the budget.

We met with Malign, who we have taken on as our tour guide. Once a policeman, he now drives a taxi. What a relief to have found an important contact on our own, someone we can trust. Malign took Nazila, Rose, and me to look at potential film locations. I asked him to drive us around the countryside so that I could get a sense of the colors, the light, and the vegetation. We saw huge termite hills scattered along the roads. The dusty red earth here settles on everything, including on the leaves of the palm trees, which looked as if they were made of red earth.

Later we went to give Bilaela half of the agreed-upon money. She complained about not receiving the other half, and insisted that we count the money in front of her, so that she could be sure we weren't cheating her! We have reached a compromise agreement with her, and among other things, she has arranged for us to shoot the "coming out" ceremony this Sunday, as well as the workshop with the traditional midwives, the one the local campaign against female genital mutilation runs on a regular basis.

If it were not for the fact that Alice is arriving tomorrow, I am certain Bilaela would not continue to work with us.

### Friday, 5 February 1993

Alice and the crew arrived today and all day I was simultaneously excited and nervous. We had to arrange for vans and cars to pick them up, along with all the equipment. I went into the arrival compound to greet them. It was wonderful to see them all, and I got such a buzz from seeing all those wonderful shiny silver boxes containing the equipment! The best part of a production for me is the shoot itself, and now, finally, it's happening.

I'd been worried about the customs process, but Malign, who has turned out to be amazingly well-connected, spoke to a friend of his who just happens to be inspector for airport customs, and he eased us through.

## SATURDAY, 6 FEBRUARY 1993

The crew had a rest day today, except for a meeting with Naz and me to discuss the shooting schedule for the week. Judy, the sound recordist, and Lorraine, the camera assistant, are the two crew members with whom I haven't worked before. Nancy Schiesari, the camerawoman, and I have worked together on two previous films, so we are somewhat familiar with each other's ways.

Had a long meeting with Alice and Deborah Matthews, who is accompanying Alice as her assistant. I am glad Deborah is here to keep Alice company, because we won't be needing Alice every day. I have arranged for Malign to be at their disposal for sightseeing.

Alice and I found some time to spend alone together, and we talked. She sounded so excited about her new love back in the U.S.! She was beaming and smiling and was quite irrepressible.

## SUNDAY, 7 FEBRUARY 1993

The first day of the shoot! It was traumatic and emotionally draining. I don't know what I could have done to have prepared myself for what we witnessed today. We filmed at the village of Dar Salamay, which we'd visited during the first week of the research period. We had arranged to come back today to film the girls who were "coming out" of the bush after having been excised two weeks ago.

Today was the day when all the women in the village were preparing for the coming-out ceremony. Traditionally, it is at this time that the initiation into adulthood takes place, with special songs and dances and chants. When girls as young as four undergo this initiation and excision, I wonder what they could possibly learn about wifely duties!

It seems the village as a whole practices excision, though not infibulation. The coming-out ceremony will be the first public appearance by the girls since their excision.

We arrived at the village with Fatima, our interpreter. While we waited in the same courtyard where we stood a week ago, when we had come to introduce ourselves, we were given a wonderful surprise. A large group of girls from about three to ten years of age, as well as a few boys, started

*The children put on a spontaneous "show" for us. Dar Salamay, The Gambia.*

playing clapping games and performing for us. A young boy with a makeshift mask on his face and a stick in his hand rushed into the courtyard with a group of boys behind him and danced, while the girls looked on and clapped. This spontaneous performance by the children of the village was a charming gift and made us feel welcome.

I was glad that I'd had the chance to film the faces of these young people. I thought about the fact that the girls in the dancing, laughing group, who hadn't already been excised, would be in the near future. I detected a sadness in the eyes of many of them.

There was an air of anticipation. Alice sat in the courtyard talking with Bilaela and looked at ease. She had enjoyed watching the children dance as much as the rest of the crew.

While we waited for the coming-out ceremony to begin, I decided I would like to do an interview with a mother of one of the recently excised girls. I wanted to know why mothers put their daughters through the terrible pain they themselves had experienced firsthand. The answer seems to be that the weight of tradition is so heavy that to protest is to risk your life, or at the very least risk banishment from your village and your community. Aminata Diop's experience is a painfully vivid illustration of this.

Fatima introduced Alice and me to a woman named Mary, whose daughter is one of the circumcised girls. Her daughter is also named Mary, so we began to refer to them as big Mary and little Mary.

Alice and I discussed the questions that we wanted to ask, and Fatima translated.

Big Mary was quite shy at first but soon warmed up.

Somehow I feel that the pointed questions Alice asked her on camera made her think about excision in a way she'd never before considered. Little Mary, who is only four years old, is the youngest in the group of excised girls. When big Mary said, in the course of the interview, that she would stop this practice if she had the power to do so, because of the pain, it was a powerful moment. I know I will end up using this interview almost in its entirety, because it offers at least a little hope in the form of a mother's acknowledgment that excision is painful and that most mothers would rather not put their daughters through it, given the choice.

By now it was very hot, and the crew went quickly through the crate of water.

Finally we were taken to the edge of the village where the ceremony was about to begin. A circle of women danced

*The tree at the edge of the village under which the "celebrations" took place. Dar Salamay, The Gambia.*

around a giant tree, which created an umbrella against the hot piercing sun.

As I walked through the circle, I saw a group of girls, their bowed heads covered in scarves, sitting with their legs stretched out in front of them. They looked totally stunned, bewildered, in shock and total despair. For a few minutes I just stared, and suddenly their expressions hit me with such force that I felt tears begin to roll down my cheeks. I quickly left the circle and grabbed Nazila and told her to get me in shape. I had to direct the crew and couldn't give in to this pain, not now.

Within minutes I was back in the circle, directing the crew. I didn't want to cut myself off from my feelings, but I had to put a hold on them so that the sadness I felt wouldn't stop me from doing what I had to do. I couldn't afford to get lost in my feelings, especially not when I thought of what the young girls sitting there must have been experiencing. But really, their feelings were unimaginable to me.

*Setting up the shot.*
*Dar Salamay,*
*The Gambia.*

Finding the balance between communicating with the people you are filming and standing back from them in order to think about the shot, the image, and the next question is a very difficult process. The first time I confronted this confusion of feelings and roles was in 1987, when I was making my video *Reframing AIDS*. I was interviewing a man, George Cant, who had AIDS and who was talking with pride and pain about his impending death, and I found it difficult not to share his pain as I struggled to focus on the next question on the sheet in front of me, as well as think about directing the camera operator.

I think of these young girls as little birds whose fragile bodies have been bashed, whose wings have been clipped before they can discover the power of their own souls and their erotic selves. They've been irrevocably wounded by traditions that cause them much pain and deny them the freedom to fly, to flourish.

The circumciser's knife had eradicated a source of fundamental pleasure for these girls. The knife had cut deep into their souls, putting out the sparkle in their eyes as the psychological and psychic scarification took root. There is no doubt that the images of these girls will continue to haunt me, and I believe viewers will be similarly affected by what we witnessed today. This thought kept me going through the next few weeks of the shoot. Because I needed to be strong and clearheaded during the making of the film, I was forced to push the pain I felt for these girls into the shadows, where it remained buried until I began the editing process. Later, when I began mourning the loss of wholeness in these girls' lives, their loss of sexual pleasure, I felt devastated and enraged. They had been robbed of something so primary. But for now, questions passed at random through my mind:

Would they ever experience the pure pleasure of a clitoral orgasm?

How many of these girls will develop infections?

How many of them will die as a result of the excision?

How many of them will experience excruciating pain every time they menstruate?

How many of them will decide not to perpetrate this mutilation on their daughters, and how many will keep the tradition going?

When will this cycle of violence and humiliation end?

The dancing seemed to go on endlessly. Women circled around and around the girls, with the midday sun beating down on us all. The two men who were allowed to be present drummed insistently, and all the time the girls sat there, looking frightened and shocked.

Most of the women who were dancing around the girls were called the barren women. Apparently every village has a barren women's group, whose members are widows, or women who have not been married or do not want to get married, as well as women who have been abandoned by their husbands because they cannot have children. Some of them were dressed as men, and at first we were confused, thinking that there were many men at the ceremony. Perhaps some of the women were lesbians.

The song that Alice had read during the shoot in California came into my mind, a ritual song of female circumcision. It was found in a book published in Paris in 1937, and translated from the French by a Dutch man named Hans

Plomp, whom Alice had met on her travels around the world.

*The mothers sing to the daughters on the day of circumcision:*
*"We used to be friends, but today I am the master, for I am a*
*man. Look—I have the knife in my hand, and I will operate*
*on you. Your clitoris, which you guard so jealously, I will cut*
*off and throw away, for today I am a man.*

*"My heart is cold. Otherwise I could not do this to you."*
*The daughters sing in response:*

*"Your words make us very scared, but we cannot escape.*
*You have been operated on as well, and you did not die. So*
*we will not die either."*

*The old women try to increase the fear of the girls:*

*"Don't be so sure of that, sisters. My heart almost breaks,*
*and I am deadly scared. Oh, if I could change myself into a*
*bird—oh, how fast I would fly away."*

*The girls answer:*

*"It's a disgrace to be afraid. If we'll die, it is a pity, but*
*we have to be courageous."*

*After the operation, the girls are humiliated and mocked*
*by the other women present. They sing:*

*"You thought nobody could overcome you. You ate by*
*yourself. You ran fast, but now you are wounded. You said*
*no one could ever stop you, but today you were held by two*
*women. You used to take care of yourself. Now somebody else*
*has to wash you and nurse you. You made love, but that's*
*impossible now. You used to piss well, but from now on you*
*will cry when you piss. You used to move so graciously, with*
*closed legs. Now you walk with your thighs apart, like a toad*
*or a mouse.*

*The victims respond:*

*"Today the knife has killed the guardian. Now he is dead. My village is unguarded. It used to be dirty, but now the guardian is gone. It will be clean."*

*The elderly ladies mock:*

*"You did not suffer as much as we did. Your clitoris is much smaller."*

*To which the young girls answer:*

*"My clitoris is just as large as yours, and if we could exchange our pain, you would find out we are suffering just as much as you did."*

*Finally, the oldest women in the ritual threaten:*

*"Don't you quarrel, or we will punish you. We will make your pain unbearable by rubbing salt in your wounds."\**

In this song, the women sing about becoming men in order to do the circumcision. Women literally abolish themselves as women and take on a male persona in order to participate in the ritual.

This ritual song also reveals how the deeply rooted patriarchy perpetuates this violence by turning women into heroes for withstanding the terrible pain of mutilation. The song and the dance symbolize the eradication of the bond between grandmother and mother, mother and daughter.

Perhaps this frantic dancing is a way for the women to numb themselves. I wondered if the dance helped them to be detached, to deny their pain. The complexity of this web of denial and distancing demonstrates women's ability to embody, embrace, and reinforce patriarchal power. Unfortunately the phenomenon of "colonizing" and oppressing one's own kind is not new or unique, nor is it rare.

_____

*From *Mœurs et Coutumes des Manjas* by A.M.I. Vergiat (Payot, 1937).

*"Little Mary" and
another girl being bathed,
"purified," at the close
of the "coming out"
(back into the village)
ceremony.*

*"Little Mary" being
reprimanded by one
of the dancers.*

Since the girls had never been told what was going to happen to them when they went to the bush, they probably didn't know what was going to be done to them next. We followed the women with our camera, to film the girls being washed in a purification ceremony. Then came a stern lecture to the girls from the circumciser, before a chicken was beheaded by one of the two male drummers present at this ceremony. (Later Alice said that the message is being spelled out very clearly to the girls, "If you don't do as we say, then you too will be beheaded." Control through subliminal terror.)

The headless chicken crawled toward the feet of the young girls and seemed in particular to go toward little Mary, and her blood-spattered feet recoiled from the touch of the dying chicken.

A woman dressed in men's clothes harshly held down little Mary's feet. Little Mary's eyes appeared woeful, terrified; she stared at the camera while being washed.

A woman came toward the camera, dancing defiantly.

I know some of these images will stay with me forever.

Finally the girls were given a special meal. They were also given gifts of beautiful clothes, including woven shawls. We were told that this was one of the only times in their lives when they would get to eat this well and would be the clear center of attention.

Then the procession, headed by the circumciser and her assistants, led the girls back to the village. They proudly carried the calabash, which holds their instruments of torture, on their heads. A return home after two weeks in the bush. I asked Nancy to shoot this with a handheld camera. I particularly wanted to capture images of the girls' feet shuffling along in a line as they were escorted back to the village by a group of women, dancing and singing. Little Mary was squeezed in between two taller girls. Many of the girls hid their faces behind their beautiful new shawls as they were taken to the village square, and made to dance with their mothers to the relentless drumbeat, which had not stopped once since early morning.

We were able to film an interview with the village circumciser. She was adorned in a white gown and turban, and wore a long, thick silver chain with several pendants. In her hands she held a red and green striped baton. Next to her sat her assistant, also dressed in white. By their feet was the calabash containing their instruments. The circumciser looked hard as steel. Alice sat to the side of them. It was a highly charged tableau. She was defensive and belligerent, refused to reveal very much. Everything is a secret, and "even if you put a knife to the girls who have been circumcised, they wouldn't talk and tell you what is cut and how it is done." When she said that she would never be in the company of women who are not circumcised and she could distinguish these women from the others, Alice confronted her: "There are many women among you now— all of us women—who are not circumcised. Did you know that, and how can you tell?" The circumciser just laughed and said, "You want to know everything."

This interview made me realize how deeply fossilized these customs and traditions are in women's psyches, and that even if there were legislation against female genital mutilation, it would not disappear overnight. A lot of work still would need to be done.

Masses of funding and resources are needed to train women health workers to go into the rural areas where the majority of mutilations are being performed, so that women

The children who were mutilated two weeks before our arrival. "Little Mary," four years old, is in the center.

||||||||||||

Waiting for the circumciser to emerge from her house. Dar Salamay, The Gambia.

||||||||||||

can be informed about the harmful health consequences of this practice. The government needs to come out strongly and publicly against it and not worry about losing votes. The Muslim leaders who are against it need to speak out all over the country and discourage the belief that it is in the Koran—many Islamic theologians argue that it isn't. There are many other strategies for change that African women continue to discuss, but as always, the excuse for inaction is that there is not enough money.

We did our final interview of the day with two girls aged twelve and fourteen, who had been circumcised two weeks before. We held this interview in the same courtyard we'd used for the interview with the circumciser, and even though she herself was no longer there, there were many other women nearby who monitored every word the girls spoke. Of course, the poor girls were terrified and said, almost in a robot fashion, that this was the happiest day of their lives and, yes, they would do it to their daughters because this is their tradition. It was so sad to see the light gone from their beautiful eyes, to see their drawn faces. In these last two weeks, they had been catapulted into adulthood with great violence.

It was a long, long day, an incredibly painful start to the shoot. We went through about twelve rolls of film (120 minutes), and the crew worked fast and well despite the emotionally draining and physically difficult circumstances. Nancy was quick in moving her camera around. She tuned in immediately to what I wanted. She is a sensitive soul as well as a brilliant cinematographer. This is why I like working with her.

That night I was jolted out of my sleep by Nazila. I'd been screaming and crying, and my body was in spasms.

All the emotions of the day were bursting out in the darkness as I dreamed that I was being circumcised, and shouted, "No, no, don't do it, please," while in the background, the drumming rang in my ears. The sound of drumming stayed with me for days and days.

MONDAY, 8 FEBRUARY 1993

Today's half-day shoot went smoothly except when Lorraine, the assistant camera operator, fainted right next to the termite hill. The sun was intensely hot, and she had been treated that morning for a twisted ankle.

As we prepared to film Alice at the termite hill, we saw a group of women dressed in their finest clothes walking in a group down the dusty red road. It was a powerful image that spoke to me of the determination of African women. The way they walked together conveyed a sense of sisterhood, friendship, and hope, and in fact it was the very image Alice had described when she'd said she wanted to be filmed with African women celebrating their strength. Luckily, some of these women were from Malign's village, and he introduced us to them. They were on their way to a naming (christening) ceremony.

I told Alice I thought a shot of her walking with them to the accompaniment of Labi Siffre's song "Something Inside So Strong" would be a great sequence for the film. Alice had introduced me to this song, which she loves, and so of course she liked the idea. "I knew you were thinking that," she said. We both recalled the night in London at the Hackney Empire when Labi Siffre had sung this song, prompting everyone in the audience to rise to their feet, singing and clapping:

## (Something Inside) So Strong

*The higher you build your barriers*
*The taller I become*
*The further you take my rights away*
*The faster I will run*
*You can deny me*
*You can decide to turn your face away*
*No matter 'cause there's . . .*

*Something Inside So Strong*
*I know that I can make it*
*Though you're doin' me wrong, so wrong*
*You thought that my pride was gone, oh no,*
*There's Something Inside So Strong*
*Something Inside So Strong*

*The more you refuse to hear my voice*
*The louder I will sing*
*You hide behind the walls of Jericho*
*Your lies will come tumbling*
*Deny my place in time*
*You squander wealth that's mine*
*My light will shine so brightly*
*It will blind you*
*Because there's . . .*
*Something Inside So Strong . . .*

*Brothers and sisters*
*When they insist we're just not good enough*
*Well we know better*

*Just look 'em in the eyes and say*
*"We're gonna do it anyway"*
*"We're gonna do it anyway"*
*There's Something Inside So Strong. . . .*

We used the 300mm lens to shoot the sequence. This long lens makes it possible to film people in close-up even from a distance. I hoped it would come out the way I had envisioned it. Sometimes I get frustrated that I am not behind the camera. I knew I'd have to wait until I saw the rushes to know what I had; until then I could only hope and pray.

Next we filmed Alice among the numerous termite hills that are scattered along the roads. Alice stroked the rugged surface of the termite hill, her bronze hand moving slowly and caressing it tenderly. This was such a sensuous image, a celebration of the clitoris and the pleasure it gives women. The significance of the termite hills had first been made clear to me in a passage quoted in *Possessing the Secret of Joy:*

*"The God Amma, it appeared, took a lump of clay, squeezed it in his hand and flung it from him, as he had done with the stars. The clay spread and fell on the north, which is the top, and from there stretched out to the south, which is the bottom, of the world, although the whole movement was horizontal. The earth lies flat, but the north is at the top. It extends east and west with separate members, like a foetus in the womb. It is a body, that is to say, a thing with members branching out from a central mass. This body, lying flat, face upwards, in a line from north to south, is feminine. Its sexual organ is an anthill, and its clitoris a termite hill. Amma, being lonely and desirous of intercourse with this creature, approached it. That*

*A termite hill, ancient symbol of the clitoris among the Dogon of Mali.*

*was the occasion of the first breach of the order of the universe. . . .*

*"At God's approach the termite hill rose up, barring the passage and displaying its masculinity. It was as strong as the organ of the stranger, and intercourse could not take place. But God is all-powerful. He cut down the termite hill, and had intercourse with the excised earth. But the original incident was destined to affect the course of things forever. . . .*

*God had further intercourse with his earth-wife, and this time without mishaps of any kind, the excision of the offending member having removed the cause of the former disorder."\**

This was part of a creation theory found among the Dogon people in northern Mali, where female genital mutilation is widely practiced.


TUESDAY, 9 FEBRUARY 1993

Happy birthday, Alice!

I had chosen a birthday card for Alice before leaving London. She has been in good humor and easy to be with on the shoot. She is the least of my worries. Her faith and confidence in me helps to keep me going through this exciting but often stressful time.

This morning we filmed a reeducation workshop for circumcisers in Banjul. Fatima and her colleagues regularly organize these workshops for the traditional midwives and birth attendants, who are usually also the circumcisers, in an attempt to persuade them to stop circumcising girls.

---

*From *Dieu d'eau: Conversations with Ogotemmêli* by Marcel Griaule.

During this workshop, they showed slides and talked about the health dangers involved in genital mutilations. They used two female dummies, the first to illustrate genitalia intact, the second to show the effects of mutilation. They used these to talk about the problems a mutilated woman experiences during childbirth. Mary, one of the health workers who is involved in organizing these workshops, told me later that few of these women have detailed knowledge of a woman's body, and rarely do the women connect the medical problems of many excised women with the excision itself. Often the women are genuinely shocked to learn these medical facts, although all I saw today was resistance to the information that was being imparted.

There were some heated discussions between some of the circumcisers and the workshop trainers. Many of the circumcisers were defensive and had solidly entrenched attitudes. I wondered why they were there if they were not prepared to listen. It turned out that they get paid to attend these workshops. Still, at least one could hope some of the information was getting through.

I interviewed two of the circumcisers separately, and they were adamant that it was a tradition that could never be eradicated, and in any case it shouldn't be. I had been told a story of a girl whose parents, only one month before, had brought her from France to be circumcised in The Gambia, and had died at the hands of one of these women. In my interview with the circumciser, I asked if she'd ever had any girls die as a result of her "operation." She looked me in the eyes and said she had never had any "accidents."

A male Muslim scholar was also present as a guest speaker. He spoke about how female genital mutilation was not an obligatory practice according to the laws of the Koran, but the circumcisers refused to listen.

Film-wise it was a difficult situation. Extremely dim light, many people talking simultaneously, and an incessant noise from the generator were only a few of the problems. Since there was no electricity, we had a generator brought in to power the audiovisual aids needed for the workshop. And then it took a few hours after the generator finally arrived to get it to work. The noise from the generator was deafening, and I'm afraid the result will be that much of what we shot today will not be usable.

Of course, we also needed everything to be translated for us, and so I found it extremely difficult to know when to turn on the camera and when to cut. Unfortunately we couldn't just keep shooting, because we simply don't have the budget for the thousands of feet of film that this would require.

When I hear that many BBC directors use a 20:1 ratio—that is, they shoot twenty minutes of film for every minute they use—I am amazed. Most independent filmmakers can afford to go only as high as 12:1, although I know filmmakers who have even less money than we do and use a 5:1 ratio!

The best parts of today's shoot were the interviews with two young girls, both fifteen, who have set up a youth committee in their school to fight female genital mutilation. They are clear and articulate, and it's good to have this youthful energy of resistance. One of them said defiantly, "I think circumcision is very wrong, and it should be abolished." This may be the generation that will make a lot of changes!

This evening I went for a run along the beach. I used Alice's tactic and told all the boys hanging around the beach hassling women as they went by that I was praying and needed to be left alone. The strategy worked.

In a way, I *was* praying, in any case, meditating in an attempt to shake the melancholy out of my body. On the horizon I saw a delirium of colors: deep oranges, reds, purples with flashes of gold, and as I ran along the shoreline, the Atlantic breeze seemed to blow the fatigue out of my mind.

WEDNESDAY, 10 FEBRUARY 1993

We left the hotel at the crack of dawn to catch the ferry from Banjul, and went upcountry to film a woman being prepared for her wedding. I had a whole sequence worked out for this footage. I wanted to intercut shots of these preparations with an interview with Dr. Kouyate or Awa, who will talk about:

. . . how a woman has to be defibulated before her wedding day by the midwife;

. . . how the husband sometimes uses a sword or a knife to cut his wife open;

. . . how it can take up to three months before the husband can penetrate his new wife;

. . . how a woman sometimes has to be taken to hospital on her wedding night to stem the severe bleeding that can be a consequence of the husband trying to force penetration;

. . . how the husband can get so frustrated with his attempts to open up the vagina that he opts for anal penetration instead.

. . . how a new bride can often be heard screaming with the pain of forced penetration, even while the wedding celebration continues.

Today we also planned to film women gardeners who have apparently created fertile green farms on previously arid land. I want to show that women here are taking control of their own lives, are creating something out of nothing, are capable of being economically self-sufficient. These women are survivors. This was an important shoot day, and will provide essential footage for the film.

The ferry ride across the Gambia River was magical. The ferry was full of women vegetable traders, women selling batik cloth, and women trading clothes. We shot some beautiful footage with Alice on the boat and also some tracking shots along the coastline as the boat moved on. The rising sun gave off an exquisite light. There was one shot of Alice gazing out across the river, and behind her were two men looking into the water, where the sunlight was reflected. They were totally absorbed and didn't mind that we filmed them. It was a perfectly composed shot. I can imagine using these shots with music by Oumou Sangare, a popular woman singer from Mali.

Nancy and I both remarked that this reminded us of Chris Marker's film *Sans Soleil (Sunless),* one of those films I had not heard of until after I made my first video, *Emergence,* when people asked me if I had been influenced by him. *Sunless* was the first film I'd seen of his, and I was stimulated and excited by the way it made visual and intellectual demands on the viewer.

I was able to use my little Hi8 video camera, which I had hoped to use a lot more; it has been impossible so far, as the shoot has been relentlessly demanding. We were a large group: the crew, the equipment van, Alice and Deborah in Malign's taxi. Bilaela made it onto the ferry at the last moment, with Fatima and Mary, another interpreter.

The land on the other side of the Gambia River was wilder, dustier, and hotter. We all followed Bilaela's Land Rover. People seem to drive at crazy speeds here, and we had to keep telling our drivers to slow down. I did not want an accident. We asked Bilaela to tell her driver to keep us in sight; we had no idea where we were going and had to follow them. We lost sight of their jeep anyway, but I told our drivers I didn't want them driving at breakneck speeds in order to catch up. Everyone on the crew was nervous, and I was worried about their safety.

As our van turned a corner we got a shock: Bilaela, her driver, and the two interpreters stood in the middle of the road, dazed, their big Land Rover in a ditch on its head, completely smashed, gasoline leaking from the tank. We got out and told them to move quickly from their car. I was afraid it would explode.

Miraculously they seemed to have escaped with only bruises, but we couldn't be sure that they hadn't broken anything, and Fatima and Mary were in shock. I thought the best thing would be for us to go back to Banjul, to hospital.

Bilaela didn't want to turn back, so we drove another thirty kilometers to the nearest health center. There was no doctor.

Bilaela insisted that we proceed with our plans for the

day. There was nothing we could do about it. Bilaela is a very determined woman and wouldn't listen to my suggestion that we turn back to Banjul.

We drove on in the scorching heat for a couple of hours and came to another health center. A nurse cleaned their cuts and bruises, while the rest of us sat under a large leafy tree.

We reached another ferry crossing and had to wait an hour for the ferry to depart. In The Gambia, there are always many river crossings to make before you reach your destination.

By midday the temperature was in the hundreds, and we did not have air-conditioning in our van. Bilaela had decided that our first stop would be the gardens, where the women gardeners were waiting for us. It was a wonderful sight, many small gardens apparently tenderly cared for. It's hard to believe what a green, fertile, fresh oasis these women have created out of the aridness around them. The faith, labor, and creativity required was nearly unimaginable. Alice loved the gardens and took to the women immediately, as she is a gardener herself.

We would have needed an hour to shoot a proper sequence here, and we didn't have that much time. Bilaela finally agreed to abandon the shoot for that day and return to Banjul.

We got back late. Lying in bed, I kept thinking how incredible it was that no one had died in that accident. I thought about what would have happened in The Gambia to the movement against female genital mutilation in the absence of Bilaela.

## Thursday, 11 February 1993

My birthday today. I woke up feeling shaky. Yesterday was a day I won't forget too soon.

Wendy called me from Canada to wish me a happy birthday. It must have cost her a fortune, but it was good to hear from her.

I had to speak to my mother today, since it's a family tradition that we speak to each other on our birthdays. When I rang, she said she had just been doing her morning prayers and was thinking about me. I felt quite tearful and missed her. I also miss my father. Sometimes I almost forget that he died six years ago. It's strange how you always think about your parents on your birthday, no matter what age you are.

I went for a half-hour massage, hoping it would help my splitting headache. Rose was kind and caring while she massaged me. At the clinic, I ran into Alice, who gave me one of her deep, deep hugs. She said she had a birthday surprise for me, that it had just arrived this morning. I wondered what it could be.

When I got back to my room I felt very weepy. Yesterday's events were catching up with me. I really wanted to speak to Shaheen, so I phoned her in London. It was good to hear her calm, loving, reassuring voice as I cried. My soul mate.

Somehow after that I felt ready to get on with the day's shoot, which ended up going very well. I am sure that had a lot to do with the fact that we had planned it all ourselves with Malign and hadn't had to depend on anyone else.

In the morning we filmed at the local market, where the

women sell a variety of goods. I want to use as background in the film shots of women going about their everyday tasks. I think it's important to work this in; it will provide a sense of the texture and diversity of these women's daily experiences.

Alice wasn't with us today, as most of the shoot consisted of general shots of the streets, the marketplace, plus a visit to Malign's village to film women in his family preparing food for a naming ceremony.

After Malign obtained the women's permission, we did tracking shots of the clothes and vegetable stalls at the market from inside the van, as our driver, Ibrahim, drove slowly around. Nancy had rigged up the camera by the window, with Ibrahim's assistance. It was a bit bumpy inside the van because of the uneven roads, so I asked Nancy to shoot at a slower speed in order to get rid of some of the jerkiness. The air-conditioning helped to keep the crew happy.

The women here dress in beautiful, colorful fabrics, which Nancy contends are a creative reflection of the women's pain. She says pain has to go somewhere, and these flamboyant colors seem to be one expression of it.

We filmed the women pounding millet in large wooden bowls, babies strapped to their backs. They were pounding to a rhythm, making music. I will add these images to my growing repertoire of images of women engaged in everyday tasks, and I hope in the end I will have created a mosaic of women's lives.

After a lunch of sandwiches and fruit, Malign took us to his village, Jambour. His sister has just had a baby, who is utterly adorable and has a full head of hair already. It was quite a new experience for me to hold a week-old baby.

Makes me broody. Here it seems mothers are not allowed out of their homes until the naming ceremony, which is usually a week or two after a baby's birth. However, they are never lonely, always surrounded by visiting relatives.

I felt much better after today's shoot. Sometimes, in documentary film, you just have to go with the flow and think through new sequences, themes, and direction, often from one day to the next. Yesterday I'd missed filming a woman being prepared for her wedding, but today I got footage of women pounding and singing and working in market stalls. A different sequence than the one I had planned, but equally important.

Returning to the hotel in the early evening, I found a large birthday cake in my room, with a card from the hotel management. Nazila had organized this, I was sure.

We celebrated all our birthdays by going to the local Gambian restaurant. Wonderful food. As soon as we sat down, Alice wished me a happy birthday and said she had written a poem for me. She began to read it aloud, and I was moved to tears. The poem was beautiful and seemed to capture perfectly our experiences here in Africa, even acknowledging the difficulties we have had in trying to make this film. By the time Alice finished reading, Deborah was tearful too, as was Alice. I felt honored to receive this memorable birthday gift.

Dinner was delicious, and Alice was really on form. She can be funny, with her sharp, even wicked, sense of humor. This is the first birthday I have had in Africa since I was eleven, a month before we left Nairobi for England, twenty-seven years ago. I went to bed reading the very funny nine-page birthday fax Shaheen had sent me.

FRIDAY, 12 FEBRUARY 1993

Happy birthday, Nazila!

So much Aquarian energy on this film. That's what is driving it forward, despite all the obstacles. All morning we had to get the umpteen letters typed up in French and faxed to the various departments of the Ministry of Culture in Dakar. The shoot in Senegal is scheduled for tomorrow, and I still don't know how we are going to get into the country.

Poor Nazila! She is so tired. She needs some time off, but I can't do these letters without having her translate them into French. We have not had one day off since we came to Africa, and it's getting to us both.

Even when there is a day when we don't have a shoot, we still have to organize the crew's food or attend to other details. Production managers, producers, directors, and researchers never have a day off on shoots like this. The two of us inhabit all these roles on this production!

I decided to hire a production assistant to help Nazila in Senegal, through a friend of Alice's. She is an African-American woman named Cheryl Williams-Nam, married to a Senegalese man. Luckily, she has some experience in film production. Let's hope it works out.

I tried to buy some postcards in the hotel today. All I could find were the usual ethnographic stereotypical images of bare-breasted African women staring into the photographer's lens, images that create exotic other, images of availability for the white male tourists. Who stops to think of the reality behind these images? Certainly not the British and German tourists who are staying in this hotel, lying on the beach all day exposing themselves to ultraviolet rays, probably indulging in colonial fantasies of an era long since gone. The wife of the new Dutch manager of the hotel shouted at one of the African waiters yesterday, "Master wants cold water." We all would have drowned her in "Master" 's cold water if we could have.

The ideological potency of images such as those depicted on postcards and in films like *Out of Africa,* which I saw on a plane journey recently, feeds the despicable treatment of the African men and women forced to earn a living in these hotels by the mostly white guests and hotel owners. *Out of Africa* is a nostalgic cinematic indulgence romanticizing the African landscapes and the devotion of African servants to white masters and mistresses, as well as European imperialism.

We have noticed many older white women tourists accompanied by young Gambian men. Interesting.

I don't think the hotel has seen the likes of our little party of women before. The workers are enthralled and delighted with Alice and Deborah, especially their dreadlocked hair. Tomorrow we leave for Senegal.


SATURDAY, 13 FEBRUARY 1993

We started out before sunrise in order to catch the ferry from Banjul. The road to Dakar lies on the other side. As soon as we crossed the border into Senegal, we felt the strong, dusty wind. The roads were full of holes, and we were forced to drive on the side path. We came across lorries that had collapsed from the weight of sacks of ground nuts. Alice, who loves ground nuts, was happy to see so many.

We arrived in Kavil around midday. I of course remembered the village as the place where we'd spent the night of horrors among the cockroaches. Our interpreter, Adelaide, was not there to meet us. Hadi and Jacques were happy to see all of us, but especially Nazila.

That morning, they had gathered together the village circumciser and a few other women to meet with us. They had waited all morning, even though we had said we wouldn't be there until afternoon. Another of those perpetual communication problems we keep having.

Jacques took us to see the circumciser in her village in the interior, about five kilometers from Kavil. It turned out that she was away in the nearest town, Kaolack, mutilating some unfortunate girls. Unbelievably, she had left a message that if we wanted to film her doing circumcisions, we should go there and meet her.

Instead, I decided to film a sequence for P. K.'s story, from Awa's book. This is the firsthand account of a woman remembering the day she was circumcised. When I'd first read this harrowing narrative, I was profoundly stirred not only by the graphic details of what had been done to her, but also by the painfully direct way in which she spoke about it.

This was like shooting a drama sequence for which I know the narrative and need the images. I already had a storyboard in my mind. We filmed the enclosure within which the circumciser lives as well as the hut where she takes the girls to be circumcised. I filmed the circumciser's mother, who was once a circumciser herself, walking into the hut, as well as two women with a child watching the swinging tin door of the hut. We did a handheld shot from the point of view of a child being walked into the hut, which I had begun to think of as the house of torture. This will be a haunting sequence, and these images will be important in bringing the story alive dramatically.

Nazila had the impossible task of explaining to people in the compound why I didn't want to film everyone. It was hard to control the hundred or so kids who had all gathered around in excitement, hoping to be filmed. There was no way Judy, the sound recordist, was going to get an atmosphere sound track.

Deborah was helpful. She distracted the children by playing the drums and dancing a few hundred yards away from where we were filming. The temperature was over a hundred, and understandably, the strain was showing on the crew.

We dropped off Alice and Deborah in a little town called Sindia, just outside Dakar. Alice's novelist friend Ayi Kwei, who is Ghanaian, lives here with his children. He is a handsome man with a beautiful smile.

It was early evening by the time we said goodbye to Ibrahim, our driver, whose even temperament, calm nature, and expertise helped us over the border crossings. Our permissions from the Ministry of Culture still hadn't come through, and I had decided to take a chance and risk going into Senegal without them. Ibrahim knew many of the border police, which proved to be a terrific help.

He had a sad look when he said goodbye to Nazila. I think he'd fallen in love with her. (On her return to Los Angeles, Nazila received a letter from him, asking her to marry him!)

We had listened to Ibrahim's tapes throughout the trip to Senegal. He'd played Oumou Sangare and Sali Sidibi, musicians from Mali whose music I had heard before and loved. On that journey between The Gambia and Senegal, I made the decision that it would be wonderful to include these two women's music in the film.

We are staying at the Hotel Independence in Dakar, where the swimming pool is on the roof and has a glass bottom; the restaurant is beneath. The thought of all those bored waiters looking up at my flaying legs while I swim is not very appealing.

Judy, Lorraine, Nancy, Nazila, and I gathered for dinner that evening. Except for Nazila, everyone talked about all the shopping they had done in The Gambia and the clothes they had gotten made. I haven't had time to buy a single present for anyone, not even a gift for Nazila's birthday. But I am treasuring the beautiful sand dollars that I collected for Shaheen on the beach in The Gambia.

I don't think Alice will like this hotel much. There is a better one on the beach, but it's far too expensive for all of us to stay there.

SUNDAY, 14 FEBRUARY 1993

It's been a day for assessment and reappraisal. I have been having doubts about the material we have shot so far, the gaps that are still there, and the shape of the overall film. There is always a point at which, during the making of any of my films, I have a minicrisis of confidence. It's something I am learning about myself and my process of filmmaking. This could happen during the shoot, like today, or in the postproduction period, when I am editing.

I had a long talk with Shaheen. I hate to think what our phone bill is going to be like, but I do need to be able to talk to her!

I also had a long discussion with Nazila, who reminded me of my original intentions. It makes such a difference to have a production manager who is familiar with my films and style. She is able to remind me what makes my documentaries uniquely mine, discuss film forms and genres as well as budget matters. I wish she lived in London! I would love to work with her again, and I know she would make an excellent producer.

I have been worried that we still don't have enough interviews with women who themselves have been mutilated but will not permit their daughters to be. It is crucial to document these voices of resistance, to show women as survivors and champions, instead of victims.

I wish I could film the women in Awa's group, the Commission for the Abolition of Sexual Mutilation. I have a feeling that a discussion with them would yield what is still missing in the film. But I can't get to these women without Awa, and she has still not given us any definite commitment for a meeting with this group or a time for an interview with her.

The footage from The Gambia may be good enough, but I won't know until I receive the report from the lab. We have been freighting our rushes back to the laboratory in London, but Sophie, in the London office, seems to have gone to New York on another shoot and not sent on the lab report. It's nerve-racking. We don't know how the rushes of the footage we've shot already have turned out. Even so, I know what we shot in The Gambia is not enough.

Spent most of the afternoon with Nazila and our local production assistant, Cheryl, trying to get the replacement tripod legs released from customs. I was told it was all paid for at customs in England, but the guys here insisted we pay them more. In the end, we were forced to give one of them a large tip in order to get our equipment and stock released.

Cheryl took us to a fishing village near Yoff, a half-hour drive from Dakar, where Africa's leading filmmaker, Ousmane Sembene, lives. It was an interesting village, almost a small town, and thrilling to me since one of the greatest living African filmmakers lives there.

Fishing is a way of life for many of the coastal people in Senegal, and I am keen to include footage of the fisherwomen here, but all the agreements and permissions have to be obtained and payments agreed on. So much setting up to do! But these images of strong women conducting their own business and being economically self-determined will be the replacement footage for the women gardeners we never got in The Gambia due to the car accident.

I am tired, but I'm relieved to be into our second week of the African shoot and to know that we have managed despite the obstacles. It's also a great relief that no one has become ill, apart from a few upset stomachs. Nancy in particular had been worried that she would contract some dreadful disease. All credit to Naz and me for checking out all the hotels and the restaurants beforehand—guinea pigs for everyone else.

I have decided not to go to Ouagadougou, Burkina Faso, for the FESPACO film festival, as Alice and I had originally planned. The week-long Pan-African film festival happens only once every two years, and it's sheer coincidence that it starts a day after our shoot finishes.

I am too, too tired. I just want to go home. Alice still intends to go, with a friend from the U.S. She remains fresh and is enjoying meeting friends here that she first met during college days in the U.S. She looks radiant.

### MONDAY, 15 FEBRUARY 1993

At last, the lab report was faxed to me this morning from the London office by Rachel Wexler. One roll has a scratch on the negative and will be unusable, but everything else is apparently fine. What a relief!

We had arranged to meet Madame Fall at 9:00 A.M. in the courtyard where we attended the PDS women's group meeting, on the outskirts of Dakar. We arrived on time to film an interview with her and the other women in her group, but no Madame Fall.

Nazila and I went down the street to find Madame Fall. It turned out she had not yet had her breakfast, so I had to decide what to do with the crew while we waited for her. Nazila stayed with Madame Fall to hurry her along.

We didn't start until a few hours later, and by this time everyone was hungry, thirsty, and generally fed up. Cheryl had been delayed in bringing the cold drinks and snacks, and it was extremely hot.

After all the waiting, the interview turned out to be a great disappointment. In our previous conversations, Madame Fall had spoken of the joys of sex, how important it was to her, how circumcision took this pleasure from women. She had laughed, had been informative, entertain-

ing, persuasive. None of this occurred in our filmed interview, and none of our promptings produced what we were
looking for. Instead, she talked about her "leader" and the
political party she belonged to.

Fortunately, the interviews with the two sisters were in-
spiring. One of the sisters said, "You cannot ever come to
terms with pain. The memory always stays, even if you are
only two years of age. A tetanus injection pales in comparison with the traumatic pain of having to sit with your legs
apart and watch them cutting your body."

The interview with the younger sister was hopeful evidence that the tradition of mothers mutilating their daughters may finally stop. "As for me," she said, "since I was
young I have always said, should I have a young daughter
she would never be mutilated. I haven't forgotten my own
experience of being mutilated."

Their spirit of resistance reminded me of what James
Baldwin once said: "The victim who articulates the situation of the victim is no longer a victim; he or she is a threat."

The older sister's daughter was around the entire morning, so I decided to film her, her tiny legs walking through
the courtyard, which was paved with old mosaic tiles. We
also got extreme close-up shots of her photogenic face just
staring into the camera. By opening the viewer's heart to
the preciousness and vulnerability of the little girls soon to
be wounded by genital mutilation, these shots will go a long
way toward explaining why we feel compelled to make the
film.

We sent the crew off to Cheryl's house for lunch, while
Nazila and I stayed with the two sisters and Madame Fall.
The sisters had prepared lunch for us. Fish again! Nazila and

I have both lost a lot of weight, and the circles under my eyes get darker each day.

This afternoon we watched a very old woman, nearly doubled over with age, being helped into the courtyard by her daughter: this was the old circumciser Madame Fall had said we should talk to for the film. She had a bundle with her, which contained all her aids and tools. Her daughter conducted the interview for us. This was because they only spoke Mandinka and Madame Fall only spoke Wolof, which presented a serious communication problem. In the end, one of the sisters spoke a bit of Mandinka and could translate the questions from Wolof.

The language chain was such that I asked the question in English, Nazila translated into French, Madame Fall translated into Wolof, one of the sisters translated into Mandinka, and then the daughter asked her mother the question. It was very sad that this woman who is really too old to be working at all is still cutting up small girls for a living. She can hardly see, can barely walk, but she is only too willing to show us her knife and talk in detail about her "work." She is the only one we've met so far who is willing to talk openly about how the circumcision is done: what is used, what is cut. At first we couldn't understand why she kept looking at the sound recordist and talking exclusively to her. It turned out Judy looks just like her niece, and she couldn't get over the resemblance.

The circumciser was very matter-of-fact when she described the process: "When I am ready with the knife, I use the water from this bottle to use on the eyes. I do it with this knife, and afterward I put some cotton wool and alcohol on it. I perform it, and afterward I spit and the hemorrhaging stops." When I asked her what she cuts, her reply made me nauseous and faint: "There is a small part that is shaped like a horn, which I cut, then I wash with salted water before applying alcohol."

It was chilling to know that only two days ago this old woman had circumcised two girls, and it was sad that not only had she apparently done it without question, she was even proud of her work. She told us: "It is this knife which has paid for my clothes and my food and my home all these years."

One of the sisters had said in her interview: "The mothers and the grandmothers, who should have been protecting them, are cutting them." It is a tragedy that the grandmother—who in many cultures, including the African, is pivotal to the community and commands great respect—is often the circumciser. She is the one who is the source of ancestral knowledge. She is the one whose presence is an absolute necessity for all rites of passage—birth, naming ceremonies, and death. *Possessing the Secret of Joy* is sensitive to this, as it is to the circumciser, M'Lissa, who is not excused but interrogated and her inner turmoil exposed.

*Can you imagine the life of the* tsunga *[circumciser] who feels? I learned not to feel. You can learn not to. In this I was like my grandmother, who became so callous people called her "I Am a Belly." She would circumcise the children and demand food immediately after; even if the child still screamed. For my mother it was a torture.*

*Then, one day, my mother had to circumcise the girls in my age group. . . . And when my turn came she tried to get away with cutting lightly. . . . What my mother started the witch doctor finished. . . . He showed no mercy. In fright and unbearable pain my body bucked under the razor-sharp stone he was cutting me with . . .*

*I could never again see myself, for the child that finally rose from the mat three months later, and dragged herself out of the initiation hut and finally home, was not the child who had been taken there. I was never to see that child again.*

*. . . I have been strong. . . . Strong and brave. Dragging my half-body wherever half a body was needed. In service to tradition, to what makes us a people. In service to the country and what makes us who we are. But who are we but torturers of children?*

Early that evening, the women began to arrive for a meeting of the PDS women's group. Alice came for this part of the shoot. We planned to film her first meeting with Awa Thiam, an encounter I thought of as a kind of historic occasion.

*Alice in a courtyard filled with women. Dakar, Senegal.*

IIIIIIIIIII

The members of the group were beautifully dressed in

oranges, greens, blues, and pinks. I am hoping this footage will show a group of women coming together, organizing around concrete political actions. Since we do not have a sound assistant, we are just going to have to get what we can, but the main point is to get the visuals and cutaways to be intercut with Awa's interview.

I also wanted to film the women dancing, since dance seems to be an integral part of the culture for women in this part of Africa. My idea is to start the film with shots of women dancing, in slow motion, full frame, the beautiful blue and orange materials moving about rhythmically.

## Tuesday, 16 February 1993

Nazila woke up with a stiff neck and looked as if she was in a lot of pain. The intensity of the work and the painful subject matter of the film are taking their toll, and Nazila and I both are struggling to keep ourselves intact. Thankfully, I haven't had problems with my back since I've been here, even though my back is my weak spot.

The crew had the morning off while Naz and I went off to the bank. That afternoon, Nazila insisted on coming to film the baobab trees, and then the fishing village, despite her pain. We don't have Alice with us today.

On our journey back from filming the surreal baobab trees, some of which are hundreds of years old, I persuaded Nazila to go back to the hotel to rest. Her pain had gotten worse; I could see it in her face.

The baobab trees are unique to Senegal, and myths about them abound. I wanted to film them because they looked mutilated; somehow they seemed to me to be symbols of the women's mutilations. We also filmed women walking near the trees with buckets on their heads, going from their villages to the nearest town. The vibrant colors of their dresses against the dusty landscape was beautiful.

We decided to take some shots of the women bartering for the fish they'd caught that day. I'd never seen so many huge fish! We got soaked by the incoming tide during the shoot, but even so, we enjoyed this time at the fishing village.

### WEDNESDAY, 17 FEBRUARY 1993

"The circumciser has her own blade, she cuts and passes from one child to the next with the same blade, soiled with blood, her hands also soiled with blood. So, obviously, if she's a carrier of AIDS and cuts herself, she can transmit it. If one of the excised children is a carrier of AIDS, she can transmit it. More and more you see HIV-positive children born of HIV-positive parents."

This is one of the chilling facts that emerged today in the interview with Dr. Henriette Kouyate at her clinic in Dakar. Dr. Kouyate is the secretary of COSEPRAT, the committee fighting against harmful traditional practices.

The interview with her was shot this morning, and is one of the most informative we have filmed to date. I liked Dr. Kouyate's integrity and her humble, powerful delivery. She is a graceful, committed, and articulate woman. She has been working against genital mutilation since 1955. Her insights about the role of tradition and about Islam were revealing.

"Nowhere in the Koran have we found it said that excision is obligatory. To prove it, in Morocco women aren't excised. Nor are Muslim women in Saudi Arabia. Wolof women are Muslims but aren't excised. As with all traditions, you must take the good and eliminate the bad while remaining true to yourself."

This incisive remark challenges the worrying and dangerous tendency in cultural nationalist movements, which in their bid to return to traditional values deny women their free agency. It is women who bear the brunt of these patriarchal traditions. For instance, women in Iran have been forced to put back the veil in the Islamic drive for a return to traditional values. While a sense of cultural nationalism is often crucial to the fight for self-determination, this should not be at the expense of half the nation's population!

In the afternoon, we met Nafi, Daniele, and Khady, three women from Awa's group, the Commission for the Abolition of Sexual Mutilation. We filmed them as a unit, talking about the aims and objectives of their organization, and the different ways the group has attempted to raise people's consciousness about genital mutilation, as well as other forms of oppression against women. For example, last year they'd organized a major conference on domestic violence, for which the press coverage was apparently excellent. As a result, many men had come forward to offer their help in stemming the all-too-common occurrence of domestic violence. This group sees female genital mutilation as only one aspect of women's oppression.

Today's shoot has yielded an abundance of important material, and in this sense I couldn't have hoped for more. I can't believe how everything I have wanted to film in

Senegal has come through despite the many obstacles. It's a good thing I refuse to take no for an answer!

We went to the most popular local restaurant for lunch and bumped into another U.S. film crew. I was pleasantly surprised to run into Manthia Diawara, whom I'd first met a couple of years ago in London. Manthia was in Senegal to coproduce a film being directed by the Kenyan writer Ngugi wa Thiong'o on Ousmane Sembene, and our crews exchanged notes, talked about exposure problems, stock, and local hangout places.

I have long admired Manthia's writings on African cinema, and so it was a particular pleasure to encounter him here in Senegal.

### THURSDAY, 18 FEBRUARY 1993

Today's experience was exciting and profoundly moving. We filmed on Gorée Island, where the House of Slaves is located. On the twenty-minute ferry trip to Gorée, we shot some beautiful, reflective shots of Alice. There were only a few tourists on the boat; the other people were mostly traders and women selling beads.

The curator of the House of Slaves, Joseph Ndiaya, was there to greet us. While the crew set up, we went into Mr. Ndiaya's office and he began to talk to us in a soothing voice about the history of the house. He said to Alice and Deborah: "I know you have come here on a pilgrimage, and it is good that you are here. You are very welcome. You could be my distant cousins or my long-lost relatives. We, each of us, are related."

Alice was very moved and cried deeply. She wrote a special message in the visitors book and saw that other distinguished visitors had been there before her. Photographs of people like Nina Simone, Danny Glover, Jesse Jackson, and the Pope were hung all over the walls, along with quotes about the legacy of slavery; there were also paintings by Mr. Ndiaya.

I set about filming the final interview with Alice. She was clear, strong, and precise. She said: "Do we care about African children, or are we like the midwife who says she doesn't hear the little girl screaming when she is cutting her? Are we expected to be deaf?"

From the beginning, Alice's clarity of purpose and thoughts about female genital mutilation have been indispensable to me and to the film. Alice has of course spent many years thinking about this subject and has already gone through many of the emotions and questionings that are just coming up for me now.

I thought back to a forum at the Institute of Contemporary Arts in London, which I'd been asked to chair. Alice, in London to promote *Possessing the Secret of Joy,* was on a panel with Efua Dorkenoo and a few other women. We had expected Alice to speak about female genital mutilation, but instead she talked about music and in particular the African music that had sustained her during the writing of the book. It is only now that I am beginning to understand how she found in music a way to express resistance and defiance, but also to lighten the burden of pain, to undergo healing.

Tomorrow is the last day of the shoot! Miraculously, we have all survived these weeks, although I confess I am ready to drop. I don't know where my energy has come from.

The day ended with Alice and me meeting up with Tracy

Chapman, who was visiting Senegal. We met in Dakar at Alice's hotel. I thought it was a rare historic opportunity to film two highly acclaimed African-American women making a pilgrimage to the House of Slaves. I also decided to ask Tracy to say something about female genital mutilation. The more people who speak out against this, the more the issue will get the attention and resources it requires. Tracy was receptive to my request, and we are going to film her tomorrow. Serendipity!

### FRIDAY, 19 FEBRUARY 1993

It's the last day of the shoot, and everyone seems happy. At last!

The interview with Tracy on Gorée Island was excellent. I don't think I have ever seen or heard her say that much in any film or interview: she is known to be camera shy and a private person. How gratifying that she agreed to be filmed for *Warrior Marks!*

Awa arrived at Alice's hotel for the interview we had arranged. Could we reschedule the interview for another day? she asked. No way, Awa. I took her hand and sat her in the chair and told her she could not go anywhere until we had done the interview. I think she quite liked my bullying.

Awa looked splendid, as always. I was touched that she wore the blue shawl I had given her. I had prepared a list of questions for Alice, which Nazila translated into French for Awa. I couldn't believe it, but I finally got the interview I wanted. The film will benefit greatly from Awa's insights and her analysis, so I guess all the angst and running around were well worth it.

*From right: Alice, Pratibha, Deborah, and crew. Gorée Island, Senegal.*

‖‖‖‖‖‖‖‖

*Bottom row, from left: Tracy Chapman, Pratibha Parmar, Alice Walker, Nazila Hedayat, and local boys at the House of Slaves.*

‖‖‖‖‖‖‖‖

*Two excised girls at the "coming out" ceremony, Dar Salamay.*

‖‖‖‖‖‖‖‖

Nazila and I spent our last night in Dakar at Awa's house, where we joined Awa, Khady, Daniele, and Nafi for a sumptuous feast. We also met her four-year-old son, who is adorable. Awa wore a stunning electric-blue outfit and was very entertaining. I think I am going to miss her. She has become so much a part of my life this last month!

That night and the next day on the plane I felt emotional, and all I wanted to do was cry with relief, hope, and pain. A multitude of emotions have been jostling for attention this last month, when all I could do was deal with the shooting schedule, the budget, interviews, and the crew's needs. Maybe now I will have the time to grieve and confront the haunting images of young wounded girls, of brutalized bodies and broken spirits, of saddened faces, the light stolen from innocent eyes.

As I leave Africa, I reflect on the footage I am taking with me, which, over the next few months, I will view again and again, weaving words and images into a film that I hope will do justice to the dignity, the pain, the torture, and the horror I have witnessed. May the film help women's voices of self-determination and resistance to be heard across the lands!

## Sculpting in Time*: Editing Warrior Marks

15 March – 2 April 1993
*Edit Weeks 1–3*

My film editor, Anna Liebschner, and I had eight weeks to edit the total footage down from eight hundred to fifty-two minutes. Working on television budgets means you also have to work on television deadlines and constraints. Ideally, I would have liked to have had a Steenbeck (a 16mm film-editing deck) in my home and to have edited the film over a period of at least six months. Instead, for the next eight

---

*Sculpting in time is a phrase used by Russian filmmaker Andrei Tarkovsky to describe the process of editing a film.

weeks Anna and I worked eight to ten hours every day, crafting and shaping the film.

Maybe it was a reflection of my growing confidence as a filmmaker, but I wanted to be led by the images and emotions as much as by what was being said. In the past, because of editing time limitations, I'd worked on rough script edits before even entering the editing room, and this would form the basis for the first assembly edit of the film. Doing a rough script edit would involve reading all the transcripts and choosing sections I wanted to have in the film. I would then arrange these in a running order that would form the backbone of the film, its themes, information, and analysis.

This time, I was adamant that I wouldn't begin that way. Instead, I read and reread the typed and translated transcripts of all the interviews and stored them in my memory bank. I wanted to approach the editing of this film organically and intuitively, working with the images as much as with the spoken words.

The week before Anna and I started working together, I sat alone in the editing room, viewing the rushes and familiarizing myself with them so deeply that patterns began to emerge. Instead of trepidation, I began to feel elation, because I could already see images that belonged together; certain sequences were already cut in my mind, and I made notes of shots that *had* to be in the film. However, I was disappointed with some of the footage. In my mind's eye, certain images had been vastly different from what I saw in front of me now on the small screen of the Steenbeck.

Alice stopped off in London for a couple of days on her way from Africa to California and I took this opportunity to show her the rushes. We both looked at them with excitement, but also with sadness. So many of the images transported us immediately back to Africa and renewed the painful memories.

One major decision I had to make within the first two weeks of editing concerned my treatment of excision via P. K.'s story, from Awa's book *Black Sisters, Speak Out*. I had wanted to dramatize this with a shoot in a studio, creating emotional vignettes that would be interspersed throughout the film. I knew I did not want a naturalistic drama, nor did I want a graphic rendition, which could become so technical that the emotions could get lost. Nor did I want to sensationalize the story or paralyze the viewers with horror.

How could I accomplish all of this and still retain the emotional integrity of the story? The film needed to be discreet but still powerful, be revealing, respectful, and never minimize the pain of the women. How could I represent the unrepresentable?

Awa's book uses "P. K."'s story to convey in relentless detail the harrowing memory of excision, a childhood nightmare related in sobering detail, moving, terrifying and unforgettable.

I had to find a way of connecting with the aesthetic feelings that were aroused in me when I read and reread this story. How could I translate onto film the interior world of a twelve-year-old girl who is being mutilated against her will? What sort of visual manifestation of this story would be appropriate? What symbolic imagery could I use to transfix the audience and do justice to this story of torture, spoken with such direct and intimate honesty? Every night for about a week, I woke up in a cold sweat, asking myself: How, how, how?

Flashes of illumination do not usually occur in a vacuum

but as part of an unconscious process of elimination and concentrated thinking. I don't remember the precise moment during which I thought: I would like to work with a dancer who could give expression to this story. I have worked with dance and dancers in my previous films, including *Sari Red, Flesh & Paper, A Place of Rage,* and *Khush.* At times I'd used dance to create a sensual, erotic charge, as in *Khush* and *Flesh & Paper,* and at other times I'd used it to celebrate women and their achievements, as in *A Place of Rage.* I'd first employed dance in a clip from an Indian movie, *Pakeeza,* which depicted a woman dancing on broken glass to express grief at her sense of patriarchal abandonment and betrayal.

For *Warrior Marks,* I thought dance could work to celebrate the sensuality of women's bodies, as well as to express the inner turmoil and devastation associated with the loss of sexual pleasure. Dance would evoke the symbolic resonance of this story, as opposed to being illustrative. But how was I going to find a dancer who could enter into the interior life of a twelve-year-old girl experiencing the horrors of sexual mutilation?

I have often marveled at the networks of chance meetings and complex associations and feelings that have brought just the right people to this film. An incredible number of "coincidences" occurred during the making of *Warrior Marks,* and sure enough, as I agonized over finding exactly the right dancer, I met Richelle.

This is how it happened. June Jordan was visiting London to promote her two new books, *Technical Difficulties* and *Haruko and Other Love Poetry.* At one of June's readings, at Brixton Town Hall, Richelle came up and greeted me. I had met her briefly at the Modern Times Bookshop in San Francisco, during a memorial for the poet Audre Lorde. Alice had taken me there a few days before we were to start our shoot in California.

Here was my dancer! I don't quite understand how these things happen, and I always end up saying, "Well, this was meant to be." I was brought up as a Hindu, and while my belief in the spiritual patterns that are "meant to be" has no rational or logical explanation, I don't dismiss these encounters as mere luck or chance or coincidence.

The dance sequence came together in a most magical way as well. I began to work with Shaheen on a set design for the dance scene. Shaheen is an architect, and when she works with me on a film, she brings to it a keen sense of design. It helps that we live together. We can talk ideas out at all hours of the day or night. Maybe it's our closeness that gives us an intuitive understanding of each other's visual conceptions, and it is Shaheen who has, over the years, encouraged me to draw and sketch and have confidence in my own ability to visualize, to be bold and take risks.

I wanted the dancer to be visually connected to the young Gambian girls we filmed on our first day in Africa, and so I decided to do a back projection, using this footage as part of the set. The rest of the design was conceived by Shaheen, and together we drew the sketches.

Two weeks into the edit period, Anna and I had already edited the first fifteen minutes. We were excited and energized. Anna, whom I had worked with before, was in great form, pushing herself as well as me to move on and on. She was as committed to creating a wonderful film as I was.

Meanwhile my body had finally decided to give up on

me. My back just gave out one morning. I couldn't move without excruciating pain. I had been back from Africa only two weeks, and in those two weeks I'd had to do the accounts, get the French interviews we had done in Senegal translated into English, read the transcripts, and prepare myself for the editing. There had been no time for rest or recuperation!

I felt that the pain I had interiorized beginning on the very first day of the shoot in The Gambia was finally surfacing, demanding to be dealt with. Nevertheless, I had to get back into the editing room with Anna, so Fola, our spirited and committed assistant editor, brought me ice packs every two hours. Later, I learned that I was not the only one feeling poorly: Deborah Matthews, Alice's assistant and friend, had also suffered from back problems after returning from Africa, while Nazila had come down with hepatitis!

4 April – 23 April 1993
*Edit Weeks 4 – 6*

Toward the beginning of April I received a letter from Awa Thiam. She was most apologetic about not having been available to us as much as she would have liked, during the shooting of our film in Senegal. She regretted that we had come at a time when she was in the last stages of preelection lobbying for the political party she belonged to. I was nicely surprised and pleased that Awa had written and was keeping in touch. She invited us to return to Senegal another time, with the film.

On a Sunday morning three weeks into the edit period, Anna and I met in the editing room and cut a sequence of images for the back projection against which I wanted Richelle to dance. The next day's shoot in the studio was incredibly stress-free, with none of the usual production headaches.

Richelle was magnificent. As the story was spoken aloud, we filmed her dancing to the music Peter Spencer had composed for the sequence.

Shaheen's team worked hard and quickly, and Jeff Baynes, a camera operator I was working with for the first time, seemed to read my mind about framing the movements, pacing the tracking shots on the dolly, and closing in on the face or going to the feet and the hands. Everyone was very moved, but also somehow disturbed by the dance.

The rushes looked fabulous when we viewed them the next day. Jeff had done a beautiful job with the lighting, Shaheen's set was exactly right, and Richelle's dancing spoke volumes.

As we began interweaving the dance sequences with the story of mutilation, it dawned on me how much this film is about bodies and body language, a celebration of women's bodies despite their mutilation. I could see in many of the images how much I wanted to celebrate the beauty of women's bodies while at the same time decrying its destruction. I wanted to evoke a sense of women loving their own bodies, reveling in their capacity to enjoy sexual pleasure through oral sexual practices. Alice returns to this idea again and again within *Possessing the Secret of Joy*.

*She was like a fleshy, succulent fruit; and when I was not with her I dreamed of the time I would next lie on my belly between her legs, my cheeks caressed by the gentle rhythms of*

*Camera operator Jeff Baynes filming dancer/choreographer Richelle.*

||||||||||||||

*Director Pratibha Parmar with Richelle.*

||||||||||||||

*her thighs. My tongue bringing us no babies, and to both of us delight. This way of loving, among her people, the greatest taboo of all.*

It was the removal of the possibility of this pleasure, really this joy, via genital mutilation, that I wanted to allude to in the film through the dancing. Mutilated women become mere vehicles for male sexual pleasure and lose the ability to control their own sexuality. Even autonomous sexual practices such as masturbation are made impossible by this loss of their sexual organs.

We were working on a stringent deadline. I was scheduled to go to California at the end of the fifth week of editing to show Alice the rough cut.

We wanted it to be as good as we could get it, and so we worked late every evening, getting it right. We went backward and forward trying to decide whether to have the rough cut transferred to video or to take it on 16mm. Finally we decided to do a rough sound dub for the 16mm print so that I wouldn't have to mix it on the Steenbeck during the screening. Before leaving, we held a screening with a few friends, including filmmaker Isaac Julien, who offered many useful comments. He was very moved by the film and excited about the way it began, with Alice's letter to me. He saw this as symbolic, a sign that women of color had begun to speak to one another across the diasporas.

I left for California feeling a little nervous, but on the whole pleased with what we had achieved in six weeks. I was relieved to leave our little edit room, which, for security reasons, had prison bars inside the windows. Worse still, it overlooked a railway line and a depressing old building

happened to be a women's prison. It was like looking out from one prison onto another. There was no relief for the eyes, no matter which way you turned.

## 25 APRIL 1993. CALIFORNIA

Alice, Deborah, Nazila, Joan (Alice's administrative assistant), Tracy, and I sat in a small edit room in downtown San Francisco for the screening. I was nervous. I normally find it nerve-racking to have screenings of rough cuts, because most people don't understand the processes involved, don't make allowances for the scratches and rawness you get on a work print.

But it didn't seem to matter: Everyone loved the film. There were tears as certain scenes evoked shared moments in Africa. We had gone on a journey together, which had profoundly changed all of us.

I had been worried about the dance sequence, since Alice had not known about it. I'd wanted it to be a surprise. Alice was worried that some of the dance shots looked sexually suggestive. She told me there is an entire pornography business that specifically depicts genitally mutilated women. I was horrified and felt nauseated by this. I don't want the film misused in this way, so I decided to take out some of the more ambiguous shots of the dancer. While in San Francisco, I grabbed the opportunity to show the rough cut of the film to Trinh T. Minh-Ha, a highly acclaimed Vietnamese-American independent filmmaker, whose films I greatly admire. I had gotten to know Minh-Ha a few years ago. Some of her early films were made in West Africa, so I was particularly keen for her to view the rough cut. Over dinner, she gave me helpful feedback.

2 MAY 1993
*On the Plane from Seattle to London*

The keynote presentation to the Northwest Lesbian and Gay Film Festival in Olympia, Washington, went very well. I had agreed to speak there because I knew I would be in California during this time. The timing was great. Visiting the students and viewing my past work with them sharpened my senses about *Warrior Marks*. They asked searching and informed questions. It always amazes me how familiar people are with my films, especially in the U.S., even though I am usually working in relative isolation, in London.

My hosts in Olympia looked after me as if I were family. For weeks I had been looking for a particular kind of music for specific sections of the film, and now this became my singular obsession. I listened to their world music collection, the best I have come across. Half an hour before leaving for the airport, I heard some music by Gretchen Langheld. It was absolutely perfect, just what I had been looking for. Later I found out that Gretchen had composed "Samburu Dance" after living in Kenya for a while. Maybe that's why I took to it immediately.

3 MAY–14 MAY 1993
*Edit Weeks 7–8*

Back in London, looking at the rough cut again after a week's break, Anna and I saw all kinds of flaws that didn't exist before. Structurally, the film wasn't yet there. I wasn't happy with it. I was also too tired and jet-lagged to think about it clearly.

A panic began to seep through me. My mind was working frantically and furiously. There were only two weeks left, and the final cut had to be ready if we were to meet all the deadlines with the neg cutters and the lab, and to stay within budget.

I went home one evening with all my little index cards, each card outlining a particular sequence or interview. The running order of the film was on these cards, which had been stuck on the edit room wall.

Too wired to sleep, I spent the night rereading the interviews, replaying all the sequences in my mind, and rearranging the whole film. At dawn I sat on the floor of my living room, surrounded by the cards and the transcripts. The panic started to subside, and I began to feel calm and confident.

I arrived early the next morning before Anna and Fola, the assistant editor, and stuck the cards back up on the wall. Anna was a bit taken aback to see that I'd made some drastic changes. It was a bit late in the schedule for this, but we decided to be bold and to go for it. Anna agreed with some of the suggestions; others we talked through, often finding a third way that worked well. What I love about filmmaking is this collaboration with so many different people at various stages of the production.

After so many weeks together Anna and I had developed an intense rapport, and now we found ourselves virtually thinking each other's thoughts. I have nurtured the director-editor relationship with Anna over my last three films. I enjoy working with her because our styles have evolved together and we've developed a kind of shorthand. Both of us are driven by a compulsion to "get it right."

The final screening in London was with Caroline Spry from Channel 4 and Debra Hauer. Neither of them had seen it before, not even the rushes. They both responded positively and offered many incisive suggestions. Debra was expecting her second child anytime that week and was in floods of tears at the scene with four-year-old Mary. I believe women all over the world will be as deeply affected.

Caroline was complimentary and wanted to see it on the big screen. She felt that using dance as I had was quite effective. Screenings with funders are usually nerve-racking, but Caroline was helpful and seemed pleased.

17 MAY – 28 MAY 1993
*Laying the Sound Track*

I felt reluctant to let go of the film, but I'd reached a point when I felt we had done our best and could do no more. It's always scary to make that final commitment to stop working. I feel vulnerable and exposed knowing that this is the version the world is going to see. It's like unveiling a part of yourself. I would have worked on it for at least another week if there had been money in the budget. But there wasn't. We were right on target.

Anna and Fola were supposed to go into the final weeks on their own, doing the track laying of the sound. Unfortunately Anna, in the third month of her pregnancy, started bleeding. Her last pregnancy had ended in a stillbirth, but even so she was amazingly considerate and waited until we had locked picture before telling me. We arranged for Cilla Beirne, an assistant editor who had worked with us on *Khush* and *A Place of Rage,* to come in and help Fola. Fola rose to the challenge, as we found out when we did the final sound mix the following week.

## 1 JUNE – 15 JUNE 1993

As soon as the first print for transmission was ready at the laboratory, I spent a day and a half transferring it onto video at Channel 4. This telecine process, where a technician and I determined the color tones from one shot to the next, was concentrated but essential work.

A major decision I had been debating throughout the editing period was whether to subtitle the film or use voice-overs. The biggest problem was that while it was technically possible to subtitle the video copy that would be used for the Channel 4 transmission, there was no place in Britain where this could be done for a 16mm print. It just wasn't done anymore. However, the lab recommended a place in Amsterdam, used by filmmakers in England. While this was expensive, I decided to go for it. I had no choice if I wanted to subtitle the 16mm print.

The interviews in The Gambia and Senegal had been conducted primarily in Wolof, Mandinka, Bambara, English, and French. In The Gambia, we'd recorded the interpreters on-site, and it was their voices that would be heard in the film.

Linda Weil-Curiel, Aminata's friend and lawyer, had done the interpreting for us during the interview with Aminata in London. This on-the-spot interpreting by someone who knew Aminata and her story intimately was entirely appropriate. Linda's commitment to publicizing Aminata's story and to fighting female genital mutilation has been steadfast.

She flew to London on a day return ticket to do a clean recording of the voice-over sections of Aminata's interview that are in the final cut of the film. Not only was it good to see her, but she also updated me on the cases she is fighting in the French courts, bringing to trial some of the perpetrators of female genital mutilation. In one case, the parents responsible for the mutilation of their one-month-old daughter are on trial. They are Gambians living in France, and had paid the excisor thirty dollars. The child's doctor had reported them to the police, and now they were refusing to release the name of the excisor, who is performing these mutilations all over Paris.

Linda thinks the parents will probably receive a prison sentence. The work she is doing is important, because every time she brings a case to trial, she forces the French government to implement its law against female genital mutilation. The more often cases are tried, the more difficult it will become for the perpetrators to get away with it. If these were white children being mutilated, there is no doubt that the Western governments would be far more vigilant and there would surely be a vociferous outcry against it.

During the final stages of postproduction, I had to make four trips to the laboratory to supervise and approve the grading of the print. In grading, one of the most important phases of printing, the technician works from one shot to the next, getting the exposure and the color tones correct on the negative of the film. I was dissatisfied with the first two graded prints, and they had to do it again until it was right.

It required all my strength and discipline not to just accept a mediocre print, since I was so tired and each trip to the lab took at least two hours. When we watched the final print, which for the first time had sound, the grader was extremely quiet throughout the screening, and at the end, was visibly affected by what he had heard. He could hardly speak. This was quite unusual, for technicians at a lab are usually concerned only about technical details, almost never about content.

Time was precious. I was starting to edit all the interviews for *Warrior Marks,* the book. These had to be ready to take with me to California, where Alice and I were planning to work together on the book. The interviews were typed up and ready on the day before I left. Jane Dibblin, a friend and also an excellent writer and editor, helped me with this.

## 16 JUNE 1993. LONDON TO SAN FRANCISCO

Shaheen and I are on the plane to San Francisco. I can't quite believe that we made it. I am so happy that Shaheen is coming with me for the Frameline Award presentation. She shares so many of my anxieties, fears, and headaches while I am working on my films. I want to share the acclaim and the celebrations with her too.

I had to phone my office from the departure lounge, to try to find out what had happened to the print that had gone from the lab to Amsterdam for subtitling. It was supposed to have come back yesterday so I could carry it with me. I trust and pray that it will arrive in the next few days, in time for my award presentation at the Castro Theater. I would like to show a clip from *Warrior Marks* along with my other work.

My time over the next few weeks will be spent working on my journal entries. I have only a few weeks to write up my journals if the book is to come out at the same time as the film. It is daunting to think that I will be writing a book with Alice Walker, who is, I can't help but remind myself, one of the most significant writers of the twentieth century and a Pulitzer Prize–winning novelist. Never did I dream that this would happen to me.

The little girl who ran barefoot on the red–earth roads of Nairobi, who climbed to her make-believe home up in the guava trees, has come a long way.

I am feeling happy, excited, apprehensive, and tired, but above all elated that the film is finally finished. There is much joy in creating work that has been inspired by passion and a desire for freedom and justice.

For me, making films is not an escape from politics but a base that sustains a political vision for democratic representation and change.

*[Artists] do not sweat and summon our best in order to rescue the killers: it is to comfort and to empower the possible victims of evil that we do tinker and daydream and revise and memorize and then impart all that we can of our inspired, our inherited, humanity.*

June Jordan
*Technical Difficulties*

The world is something that is made by human beings, and it can be remade by each and every one of us. Let us remake it with tenderness and with love. It is with that hope and in that faith that I continue to do what I do.

*Thank you, Alice!*

HERMANA DEL ALMA : SISTER OF THE SOUL

# PART THREE

# INTERVIEWS

# Alice Walker and Efua Dorkenoo

*Efua Dorkenoo is the director of FORWARD International, an activist and educational organization that works to eradicate harmful traditional practices, primarily female genital mutilation. She wrote to me after reading my book,* Possessing the Secret of Joy. *We met in October 1992, in London, at her office at the Africa Centre.*

AW: *Efua, I'm very interested to know about your work with FOR-WARD. Campaigning against female genital mutilation seems such a huge task, and yet I know that you manage to do all that you do from a very small space in the Africa Centre in London. I would like to know something about the beginnings of FORWARD and your interest in campaigning against genital mutilation.*

ED: I sometimes wake up and feel amazed that what started as such a tiny thing has grown to this size. It started when I worked with a human rights organization on women's rights issues in Africa. And stuff on genital mutilation kept coming to my desk. At the time, there was a lot being written on genital mutilation, but it was all very clinical. And I felt something went into my heart that I ought to take this on. A white woman who coauthored the Minority Rights Group Report said to me, "If I was an African woman, I would take this as the issue and work on it. But I am not an African woman, and I can only do so much."

And so I started working on it, trying to find the campaigners who are in Africa. At a certain point I went on a mission to Africa to find out what was happening and to distribute material. And that's how it started. But after two years, the project was finished; the funding bodies gave it two years, and that was the end of it. And I thought: My God, in those two years I've managed to go to the United Nations to give the first statement to the UN Human Rights Commission, and it started so many activities.

In the same period, we had two books published by African women. Previously, it had appeared as if Africans were not doing anything about it. Yet when I went to Africa, I realized that there were people working quietly, but their work is not known, they don't have a voice. So two books came out of my work. One was by a sister in Somalia who wrote *Sisters in Affliction*. And the second one was by Asma El Dareer from the Sudan, who wrote *Woman, Why Do You Weep?* And I literally went round to the publishers and said, "Please, would you publish this?"

After two years I thought that as far as the project was concerned, the money was over and that was it. But I couldn't leave it. It was like what you said—you came across this issue, and you couldn't leave it. I just could not leave the whole thing. So because there was no scope for me to then work in that organization, I decided I would start doing other work. I wanted to not only put the issue on agendas; I wanted to work at grass roots level. And FORWARD started from my home, it really did start from my home, because by that time people had been coming to me as a resource person, to tell them more about it. And so I started from my home with lectures and pamphlets. That's how it began, in 1983.

AW: *What kind of toll has this taken on your life? How has it affected your life? I know it's such a painful and exhausting subject. I just wonder how you manage to keep going; what gives you the energy to continue?*

ED: The interesting thing is that I actually found liberation in this work. Do you know, the subject opened my mind to different levels of women's position in society. So for me that has been the liberating experience.

But my whole life and my family life were disrupted. You know, first of all, as a black woman living here, I did not have access to resources. In Britain, the history of doing things on a charitable basis is about middle-class people who can afford to do this work without any toll, economically, on their family. So that was one major issue. And as a result of this, my first marriage broke down, because my husband just could not take it any longer. He said, "Why are you going out and working twenty-four hours a day and then not bringing any resources into the family?"

And it has also taken a toll on my children, I leave them quite a lot.

AW: *How old are your children?*

ED: I have a fifteen-year-old and a twelve-year-old. At one point the work was so much, and I didn't have time, so my mother came and took them to Ghana. My older one is still in Ghana now.

AW: *And they're boys?*

ED: They're boys.

AW: *Aha, but they understand?*

ED: They know female genital mutilation inside out! The younger one says to me that if he was in my position, he would just go and give orders that if you do it I'm going to take you to court. The children gave me the idea for a little booklet, *Tradition! Tradition!* I was experimenting with how we can work with women to raise their consciousness about their position in society. Genital mutilation is basically a social practice, with a health consequence. Generally, people have been dealing with it on the health side, and my impression is that if you don't get to the roots, which is the social meaning of it, we will never be able to deal with it. So I was experimenting. And the first thing I did was, I had a group of women write a drama, because I realized that in Africa, we are dramatists. Even in the marketplace, young people peddling, they will come to you and say, "Madam, this thing is very beautiful—for you, and for you alone, ten dollars, you see?" Lives are drama. And I said to myself, What are we doing? Just focusing on health education, this is bad.

AW: *We're boring them.*

ED: That's it. So get them to write their own story, and then we will put it onstage, and we'll dramatize this for more women. And it will be therapeutic, and also, it would get them into discussion on a different level. So I worked with the women for three months.

The story was about their lives. Women in Britain come over from their country in Africa, as a bride, brought from the village, to come and marry a man here, in an arranged marriage. They are infibulated, and on their wedding day they have to be defibulated—opened up. And of course the medical profession here does not understand this issue, so they cannot go to British doctors and say, Could you open me up? The doctors would just freak out. So the women do it themselves. And in order to do it without the neighbors hearing about it, they have to put on the TV, anything to muffle the noise. And the other women hold you down to do this. And this was the story we wrote.

But when we finished, and we were ready to put it on, they said to me, "Efua, if we put this drama on, we will be killed." And I had to listen to the women. Women can be killed. For one week I couldn't sleep. I said, We still have to have that drama. And the idea of using symbolism, the story, the tradition, came to my mind. Why don't we use a symbol, make it funny, you see. A society where all women have one leg amputated—the symbol of the mutilation, which is acceptable, because you are not talking about the genitals, which is taboo. And we used Mother Earth, who is like an outsider, like you, Alice, in some ways. Not coming from a particular community but trying to understand this tradition. So through the eyes of Mother Earth, we go through history and ask: How did it start? It started a long time ago, about three thousand years. But why did it start? In our story, men and women have to dance in a competition, and that is addressing the gender question. And we get to different parts of the issue but in such a simplified way the audience can laugh. And then later on we can sit down and discuss the origins, including children's human rights. In our drama, Mother Earth imagined them crying and screaming with pain.

AW: *I like your play* Tradition! Tradition! *very much, because it is good to have people understand that after a clitoridectomy you are different, as you would be after having your leg amputated. And I think many people who are brought up to think that the genitals are not important can begin to see, from what you have presented, that you are saying that if you missed your leg, you would also miss your clitoris. So they have a chance to see that at least you, the writer, value the parts of the body, no matter where they are.*

*I wanted to ask you a question that's been on my mind a lot, and that is about your parents, both of them: how they respond to your work and what you've talked about with them concerning what you do.*

ED: When I started doing this work I had already lost my father. But my mother, I think I inherited a lot of energy from her. She encouraged me, and she felt that it was as if I had a mission to do this work. When I didn't have time for my children, for example, she said, "Let me take the children so you can do this work." Sadly, I lost her last year. But even up to the time that she was about to die, she gave me her blessing for this work, assuring me that I should continue, I should not leave it.

One of the issues which always used to come up is whether I would have been better off pursuing my career, taking a big job somewhere else, rather than working in this field, where resources happen to be very scant. And she always felt that this is God's work and that I should do it. She even acted in the play—she sang. When we performed *Tradition! Tradition!* here in London, she came and took the role of Mother Earth and sang in it, something really spiritual.

AW:  *I'm really glad to hear that, because I think we get our power to act in the world through our mothers, often. And it's very good to feel that we are backed up by this kind of power, concern, and love. And that our mothers understand what we do, this is really very necessary.*

*I know that different people have gone to Africa to try to address genital mutilation. Have they gone about it in the wrong way? Why hasn't it been more successful?*

ED:  I think it's very complex. From my perspective, I think that development only works if you can energize people at grass roots level to feel they want to do things for themselves. And in most of the campaigns I've seen, people go in there and give a lecture to village people and come out. It doesn't work. It's like the people are objects.

AW: *In Kenya, in the old days, before independence, there was a very strong campaign by colonialists to end genital mutilation. And the people resisted because they saw this as another colonialist ploy to usurp and destroy their traditions. So they reinforced genital mutilation even more strongly, and they had the support of Kenyatta. Do you think that part of the reason there has been a real resistance to change is that people see it as a kind of arrogance on the part of the colonialists, the Europeans, who have come into Africa to exploit the resources of the continent and to change the way of the people, so that the people will be more exploitable by Europeans?*

ED: I think there are two sides to it. Most Africans aren't aware that there was an attempt, particularly with the missionaries, to make us see everything we did, our own culture, as primitive, bad, barbaric. They used ploys like: "Before you can go to school, you've got to become a Christian." Most of the schools were missionary schools. But I think it's more than that. Because Kenyatta was a politician, he realized this tradition was a key issue he could use as a political tool, to mobilize the people against the colonialists. And even now we find that in our campaign here, you always have opportunists. They're looking for an issue which is of sensitivity to the people. So the moment we raise the question of genital mutilation, they then mobilize people against us and say, "Look, this is your culture; people are coming in to try to change it."

The second issue has to do with the control of women's

sexuality. In any society, once you start raising issues of gender, there's going to be resistance. In terms of our nationalist struggle, we're trying to break free from the wider oppression of racism. But our women are often confused, because we are told that the most important issue to focus on is the nationalist struggle and that once we free ourselves from colonial domination, everything will be OK, and *then* we can address the women's question. When you delve into the subject, it's not just the cutting of a woman's genitals; it's also the symbolic power of it: it has implications for her psychology and character development. And therefore male-dominated society sees any attempt to change it as a threat.

AW: *Right—to their control and to their power.*

ED: That's it, that's it, you see. That's why it's so complicated.

AW: *It's like in the old days, when we were enslaved in the United States: If anyone came onto a plantation and tried to help the slave in any way, the master said, You can't, because I have control over the slave, and there is nothing I will let you do, because I basically own this person, and I control this person. Whether I flog this person, whether I brand this person, is none of your business, because this person belongs to me.*

*Now, my question is: Do you think women, by now, see through that nationalist argument? Because what has always happened is that we have stood by the men. We have said, Of course we will fight the colonialists, we will fight the plantation owner, we will fight the struggle collectively as black people. But then, when there has been some gain, and black men have risen to a position of dominance in wherever the territory is, women are once again relegated to the bottom. And all of the harmful practices that had been going on against them before continue and are often reinforced and made even worse. This has happened all over the world. In Algeria, women are back in the veil. In different places in the world, we can see that being loyal to the struggle has rarely meant that at the struggle's end we stand with our full integrity, next to the men.*

*So how do you think women are feeling about this? Do they feel that they have to continue to be loyal to what men say is the agenda? Or are they ready to say: "This hurts me; this undermines the health of the country, the health of the nation, and the health of the continent. This is as important as any other thing being discussed in Africa. And we are ready to address it." Do you think this is a wave of thought which is emerging?*

ED: Not in the context of female genital mutilation, although there are a few women who think like that at grass roots level. Amongst the intellectual core, or even amongst African women activists generally, they may not see this as an issue—that is our major problem. You see, there's a contradiction where women are really active and involved in many nationalist struggles. And I think one of the reasons is that this has been a taboo and it's not been open to that depth of discussion and debate, which would bring about more interest and thought in that direction. There has been a tendency to focus on educating the masses about the bad effects of the practice. There has not been an intellectual analysis and debate within Africa on this.

But there is hope, because I find that among the younger generation there are some amazing young women. They are analyzing, and I think that there is hope for the next generation who have started questioning. I think black women who live out of Africa are going to have a lot of impact on Africa. At a conference we had here three months ago, there was a Somali woman from Canada, slightly younger than ourselves, and she was linking this to torture. She said it should be in the UN Convention on Torture. And when we talked about victims, she said, "No, we are not victims, we are survivors." And that was wonderful to hear.

AW: *As a woman, sexual pleasure is very important to me. I remember when I started researching genital mutilation, I was very depressed to think that women had lost this pleasure—and in fact might not even know that pleasure was involved in making love or in having intact genital organs. And then I was reading Hanny Lightfoot-Klein's work. She did research in the Sudan and discovered that many women do, in fact, retain a capacity to enjoy sexual intercourse and take pleasure in making love. This made me very happy, because it pointed to something that I think is very crucial to remember, and that is that even though women may be genitally damaged, and have a very hard struggle to relate sexually to their partners, many of them manage to triumph in this area.*

*This is not the only area in which women who have been mutilated triumph. There are women who fall by the wayside, women who do crack up, women who don't make it. But there are also women who manage to find a kind of balance again in life, and they continue. I wonder if you could talk about this aspect of not being victims, of being survivors—and even going beyond being mere survivors and actually blossoming as human beings.*

ED: I think this is quite critical, because here in the West, when you talk about genital mutilation, people just imagine all women are totally destroyed. Although women have been through all this trauma, there are so many of them who are strong women doing wonderful things for the community, looking after families, doing anything and everything. You see women who have been stitched and cut and restitched, and yet you see them laughing, strong. And that is the hope for the future, within this stream of survivors, that's where things are going to change and happen.

AW: *Yes, I think it's important for young women to whom this has been done, that we should always reinforce the reality that this is not the end. Even though you have been injured, the injury is not the end of your life. Life is large, no matter how small people try to make you think it is.*

*Would you know if there are power bases for women in Africa that can be approached? And what would you anticipate the response of these power bases would be to you and your work?*

ED: We have to learn to understand that women in a male-dominated society still come together. Through rituals, they build sisterhood—sisterhood which lasts for a long time. And these rituals are mixed with the magical, so it could be said that they are part of the control of women, that the women who organize them can be seen as custodians of male power because they prepare the young women to fit into particular roles in society which are acceptable to men. But the involvement of women in organizing these magical rituals is very complex. It is a wonderful example of a woman's power base—and we should be working with them. Instead of just teaching how to be a good wife, they could be teaching women about their whole being. There is great potential in the powerful networks of women which exist in Africa. Because when girls are initiated together, they are sisters for life. Wherever you are in the world and you are in trouble, one of those sisters will come to your aid. The equivalent in the West must be the Masonic lodges, and you see the power of the Masonic lodges. So we don't want to just go in and destroy this unique form of power and sisterhood.

AW: *How do you envision the future? What would you like to do in your work? I know that there is now FORWARD International and that you are the director. This is an awful lot of responsibility. It couldn't be in better hands, and you have so much energy and vision, but it is still a great deal of work. In five years, where would you like FORWARD to be?*

ED: I would like to see FORWARD working in all countries, particularly in Africa but also in Western countries. I see my role more as a catalyst, to bring the right people together. What Africa needs now is a women's health movement. Not just family planning, which is the fashion now. The West decides suddenly that family planning is the issue, and a lot of resources are put into that. To me, women's health means paying attention to the whole woman, her physical, psychological, and spiritual well-being.

If we energize women with energy and vision, they can create the conditions for women to come together to talk about this. It could be a women's refuge. I hope women will find new modes of working with each other, giving this information back to the communities. But however we do it, we have to work through a cultural medium for women to find a new image of themselves. You see, genital mutilation is mixed up with how I see myself as a woman. Every society decides who is a woman. In British society, for example, to be a woman is to be thin.

AW: *I have a perfect slogan: To be a woman is to be healthy.*

ED: That's it. That's it. It's very simple. And women must feel that we are working for and with them, as opposed to against them.

AW: *Or condescendingly telling them how they should do it.*

ED: That's it, that's it.

# Alice Walker and Aminata Diop

*I learned of Aminata's situation from an item in the newspaper that described her flight from her home in Mali, in an attempt to escape genital mutilation, to Paris, where she was seeking political and religious asylum.*

AW: *Aminata, did you tell your mother you did not wish to be excised?*

AD:  Yes.

AW: *And what was her response?*

AD: She was very upset. She said, I've given birth to you, I've been excised, so why are you refusing? She said she was ashamed and couldn't go to the mosque. She felt it was very shaming and cried for days. She wasn't at all happy that I refused. She tried to get me to abandon my plans. She advised me to submit to it, as she had done, but I couldn't accept that.

AW: *Do you think your mother was unable to understand because of her religion?*

AD: Yes, that's what really hurt her, because my mother thought it was a religious necessity to excise her children. She thought the Koran demanded that parents have their daughters excised so they will be clean and be good Muslims. But my mother didn't really try to understand me—she was just so shocked by it. My father believed my mother helped me to flee, though she never helped me. Nobody, no one in my family, ever helped me.

AW: *But it is not in the Koran, isn't that so?*

AD: No, it's not in the Koran, but I got to know that only when I arrived here in Europe, because Africans believe it is a religious necessity, that if you are not excised you are not clean. When I told my fiancé I didn't want to be excised, he said he would not marry a woman who is unclean and who could not be a good Muslim. Africans believe that there is no point in going to the mosque or praying if you are not excised, because you can never be a proper Muslim.

AW: *When you were a little girl, what did you think excision was?*

AD: My people would tell us, "We are going to wash your hands, and afterwards you will have some food." People of other languages say, "You will sit on a piece of iron and then have food." But of course we had no idea that something was to be cut off.

AW: *Do you think that your father had any understanding of the pain involved in this procedure?*

AD: Yes. My father must have known it was painful, because he knew they had to cut something off. There are young women who can't stand the pain, and he heard some of the young girls—the pain, their shouts—so he knew. But he doesn't even try to understand—he thought it was his duty and his right to do it to his daughter. But my mother said a woman has to go through three ordeals in life: we must go through excision, marriage, and giving birth. Excision is a woman's destiny.

According to tradition, the husband should have prolonged and repeated intercourse with her during eight days. This "work" is in order to "make" an opening by preventing the scar from closing again. During these eight days, the woman remains lying down and moves as little as possible in order to keep the wound open. The morning after the wedding night, the husband puts his bloody dagger on his shoulder and makes the rounds in order to obtain general admiration.

—Jacques Lantier
*La Cité Magique*

AW: *In many cultures, excision or infibulation would occur very early in life, say at the age of eight or nine or sometimes even earlier. You were able to avoid it for a long time. Why was this?*

AD: Among my people, girls are not excised before they have breasts, before they develop and are mature. I left my village to go to the capital when I was twelve, and up to that time my father hadn't asked me to be excised. There is no special age for this, but families have to support a whole group of girls being excised together. In my group we were twenty-six girls to be excised, and everybody wants it to be a feast, which means the families have to pay a lot of money for food and celebrations.

When I was twelve, I left my village and went to live in the capital. After that, when they asked me to go through the excision with the other girls, I made an excuse that I was ill, because I'd already asked how they did the excision. I knew they cut something off, that it was very painful, and that some of the girls were badly traumatized. So I knew how painful it was. And I had a friend whom I loved a lot who died as a result of the excision. She was excised on a Thursday, and the following Monday she died. That's when I decided it would never happen to me.

AW: *Good for you. Did you explain to your fiancé that you were afraid, and did he understand what excision really is?*

AD: When I told my father I didn't want to submit to the excision, that I felt fine as I was and was no longer a virgin anyway, he did not accept it, he hit me, and lots of people gathered around to watch, as I was crying out. So I tried to explain to my fiancé when he came at night to see me; I hoped that as my future husband he would understand. I explained to my fiancé that I was frightened; I told him how painful it would be and that my friend had died and I might die, too. There was no reason for me to undergo the excision. He didn't try to understand; he said he could not accept it either, because his family has the same traditions. He said he could not be proud of a woman who would be dirty, who would be like a whore, and he would be ashamed. He let me down; he never tried to help me. He abandoned me. He not once came to look for me when I fled to the city. He reacted just like my parents.

AW: *Could you tell me what happened to your mother after you left home?*

AD: When I left home and came to France, after several months I learned that my mother had been chased from the family home. My father thought that she had helped me to get away, so he divorced her, and now she lives with her brother. My father thought that only a bad mother could have given birth to such a bad girl. So he divorced her because of me.

AW: *Do you think you will ever go back to your village or even to your homeland?*

AD: As long as my father is alive, I shall not be able to go back to the village or to the area. Maybe after, when my mother understands, I may go and visit her in the capital but not in the village.

AW: *What would it take for your mother to understand, do you think?*

AD: It's difficult, because she is suffering herself as a result, and she believes it is something shameful. The day my people understand that excision is not good, and it is forbidden to carry out excision, then maybe my mother will understand that I was right. But right now, because of the circles she moves in, she isn't ready to understand.

AW: *How old are your parents?*

AD: My mother is not yet fifty. My father was born maybe 1931 or '32; he may be seventy-five or so [*sic*].

AW: *And you?*

AD: I shall be twenty-four in December.

AW: *What do you miss most from your life in Mali?*

AD: I miss my mother and my friends—but most of all my mother. I know she suffers, and it has a hold on my life. I hate my father's guts. He and my aunt, too. They chased me away from the village.

AW: *How did you find the courage to travel so far, all the way from Mali to Paris? How did you find the courage and the strength to do this?*

AD: The day I decided not to be excised, I had not the faintest idea what would happen to me. I had no family left, only my friends. The only one to help me was Mrs. Bas. She was the eldest sister of a school friend of mine. She helped me. I tried to commit suicide, and she said, "No, you must not because it's a weakness. You must be strong." She worked for the Sabina Company, and she sent me over to them. A friend of hers took me in charge and drove me to France.

AW: *Are there other people from Mali in Paris? Have they been receptive, have they been understanding and supportive of your effort, or have they been hostile?*

AD: In the beginning, people were helpful. One of the women went to visit my mother. But now they have become very hostile, and they say a lot of evil things about me. Even the lady who visited my mother said awful things about me, like that I am a prostitute. And so my mother believed it. They said, "You should have obeyed your parents" and "You have to submit to what your parents tell you." So now they reject me completely.

AW: *What would you like to do with the rest of your life? Do you have dreams and fantasies about how you would like to live, what you would like to do?*

AD: I would like to go back to Mali—it is my aim—and I want to fight against this practice. I want to help my sisters who are under the same threat. I want to help them to avoid being circumcised.

AW: *I promise to do everything I can to make your mother understand everything you have done. Please know that I am so very deeply moved because of what you have done. I will never forget you.*

# Alice Walker and Linda Weil-Curiel

*Linda Weil-Curiel is an attorney in Paris and a friend to Aminata Diop. I learned of her work against genital mutilation also from a newspaper clipping.*

AW: *You have been a friend to Aminata Diop since she arrived in Paris. How did that come about?*

LWC: Aminata asked for the status of a refugee in France. She was sent to me by a friend of hers, Dr. Boute, an elderly lady who was a gynecologist. She'd had to sign a certificate for Aminata, showing that she has not been circumcised, and she understood that Aminata would need legal help. In France, I am known to fight excision, at least the excisions that are being performed in France by immigrants. So this is how I came to know Aminata. And we became friends.

AW: *I want to know also if your work usually entails legal counsel in defense of small children and whether it is a departure to have a case involving an adult who has come to France fleeing excision.*

LWC: Yes; Aminata's case is the exception. Normally, I do ordinary legal cases, but my feminist leanings made me consider the destiny of children who have been mutilated. I decided it was unfair, and nobody seemed to care. So on behalf of a feminist association, I decided that it was time to speak out for those children. And this is how my reputation built up and how Aminata came to me—otherwise

I wonder how I would have met such a nice and noble person.

AW: *What was the response of the French authorities to Aminata's request for asylum?*

LWC: They admitted—and it is a precedent, thanks to Aminata—that excision is a persecution under the terms of the Geneva Convention and that a woman who flees her country because of excision could be granted refugee status. But for Aminata, they said the file was not complete and she had made mistakes in her story and that therefore they would not grant her asylum. But there's another reason. The real reason is a political reason. France has many immigrants, and politicians believed that if asylum was granted to Aminata, then half of Africa would claim that they are entitled to come to France.

So this is the main reason why refugee status was not granted to Aminata. But you must know that Aminata was believed when she told her story. Because immediately after the trial, the French administration granted Aminata her papers, so she can stay legally in France, she has a working permit. But I would not let it go at that. So we decided, Aminata and I, that we should ask the Supreme Court to reconsider her demand. And we are still waiting for the decision.

AW: *I also wanted to ask you about the cases that resulted in a recent trial in France concerning a Gambian woman who took her children to the excisor. And would you tell us also about the history of excision of babies in France and the consequences?*

LWC: Unfortunately the Gambian case is not the only case. It all began when a baby girl died of her excision. She bled for three nights and days, and her parents, knowing that what they had done was illegal, had not taken the baby to hospital. When they decided to go to hospital, it was too late, and when the doctors discovered the reason for the death, they were absolutely horrified. And of course they had to alert the police.

People started talking, and other cases were brought up, because many babies are taken to hospital after excision, suffering from severe hemorrhaging. Many babies were saved, but nonetheless it is an offense to do such violence to the body and the mind of the child. So as soon as the prosecutor opened the case, I intervened as an advocate for the child on behalf of a feminist association.

We thought it was unfair to leave it only to the judges and the prosecutor to discuss. Nobody would have brought up the real question, which is: Why hurt a child? The hurt that is caused has a meaning, and it is for us women to show that meaning and to say how intolerable it is. So we had many cases, unfortunately, to deal with. Lately we caught the excisor, who was sentenced to five years' imprisonment. The parents had the benefit of extenuating circumstances and got a suspended sentence. In my opinion, it is rather unfair, because the parents are the real culprits. They know they are going to hurt the child, and they nonetheless take the child to the excisors, to the knife. They help. Do you know that the mothers hold the child on their lap, opening the legs, while the excisors cut the flesh, and they hear the child cry, and yet they do it, and there is no excuse, ever, for such a deed.

AW: *They claim to love the child, as well.*

LWC: They do, and I do not allow such a thing to be said in court.

AW: *The excisor, what did she have to say for herself?*

LWC: She kept silent. She refused to talk. But her eyes talked.

AW: *What did they say?*

LWC: "It's none of your business." But it *is* our business.

AW: *Of course it is.*

LWC: And we have to fight. We need Aminata, we need people like Aminata to help us fight. Parents are always excused for what they do to their children. So when I read your book, I was very fearful. Because each time I turned the page, I was wondering: When will the excuse for the parents come? And I am very, very happy to tell you I never found that excuse, and I thank you for it.

# Pratibha Parmar and Alice Walker

*I wanted to do this interview, at Alice's home in northern California, to establish the beginning of her journey with the film* Warrior Marks *and with the issue of female genital mutilation. We also filmed Alice telling her own story of being wounded as a child.*

PP: *Alice, why did you decide to write about female genital mutilation? You said that it was your visual mutilation that led you to think about the subject.*

AW: It helped me to see the ways in which women are rather routinely mutilated in most parts of the world and how people tend to think of the pain done to women as somehow less than pain done to men. It's often overlooked, or they think: Well, women can take it. You can see this in attitudes toward childbirth, which is incredibly painful yet is really minimized. It's a heroic thing that women are doing, bringing a child into the world—it's excruciatingly painful and yet people tend to think of it as something that's just routine: "It's a woman—she can stand it."

My own visual mutilation occurred when I was eight, and it led me to a place of great isolation in my family and in my community and to a feeling of being oppressed. It was never really explained to me, nor was there sufficient comfort given to me as a child. And I see this mirrored in the rather callous way that little girls are taken to be mutilated. You take a little child off and tell her she is going to visit her grandmother. On the way, you divert her attention from the trip to the grandmother's, and instead you hold her down and cut off her clitoris and other parts of her genitalia and leave her to heal from this as best she can. Everybody else is making merry. She is the only one crying. But somehow you don't care, you don't show sensitivity to this child's pain. I made a very strong connection with that. I was able to feel for the child, while understanding that the adults thought that they were doing the right thing.

PP: *When did you first hear about female genital mutilation?*

AW: At the time I was a student and a Westerner, and I didn't understand it. I couldn't imagine, for instance, what it was that they would remove. I mean, what is there to remove? Since obviously there isn't anything removable, I put it out of my mind for a long time. I was reading Kenyatta, who was the great liberator of Kenya, after the Mau Mau struggle. And he was saying that no Kenyan man would marry a woman who had not been mutilated. They didn't call it "mutilation," of course, but rather circumcision.

This presented a great conflict, because I was very much an admirer of Kenyatta's and very happy that he had been let out of prison—his case is very similar to Nelson Mandela's. And yet this seemed to me, even then, very strange. After all, there were many women in the struggle for liberation. And if you have women attempting to liberate a country and they can't walk, you have a problem.

PP: *How long did it take you to write your book* Possessing the Secret of Joy?

AW: To write it, I went to Mexico. I needed to be in a Third World country, where I could feel more clearly what it would be like to have a major operation without anesthetics or antiseptics, because that is what happens to little girls when they are genitally mutilated. It shows such contempt for a child's body and such contempt for the clitoris. The actual writing of the book took a year. But it took me twenty-five years since I first heard about female genital mutilation to know how to approach it. To understand what it means to all of us in the world, that you can have this kind of silencing of the pain of millions of women, over maybe six thousand years. I have talked about this in many countries, and even in Africa people stand up and say, "I have never heard of this. Are you sure that this happens?" And we are literally talking about something which has been performed on that person's mother, that person's aunt, that person's sister, that person's child, and they don't know a thing about it. It's remarkable.

PP: *What were the kinds of emotions that you felt when you were writing the book and really thinking about what this meant for women who had been genitally mutilated?*

AW: I felt a great deal of rage, a very clear burning forceful anger and rage. Because it is just unacceptable that people routinely torture children, betray the trust of children, and cause them to spend their entire lives in shame, embarrassment, and pain. If you can't be angry about the assault on the body of a defenseless child, what can you be angry about? So I felt very angry and yet increasingly clear in myself, because I believe in the power of the word to change things. I was conscious of twenty-five years of thought, growth, and preparation, and needing to take this on in the best way that I could. But I knew I could undertake this and do it well—and that was a great feeling.

PP: *What kind of responses have you had to the book and to the fact that you have written a book about a taboo subject?*

AW: Overwhelmingly positive and even joyous—especially from African and Asian women. I think women feel that, with the raising of consciousness, there really is a chance to change reality. Wherever I have been, I have had town hall meetings attended by hundreds of people—in Amsterdam, in London, in New York, in Atlanta. Women came from many parts of the world. Some of them had been mutilated, some escaped it, some of them have stories to tell about how their fathers wouldn't let it happen to them or their mothers refused to let it happen.

There are people who think that to speak about this is to stick your nose into somebody else's affairs, somebody else's culture. But there is a difference between torture and culture. I maintain that culture is not child abuse, it is not battering. People customarily do these things just as they customarily enslaved people, but slavery is not culture, nor is mutilation.

And sometimes you have to take a political or moral stand, whether it's your own or someone else's culture. Think of how happy we were, enslaved in Georgia and Alabama and Mississippi, when those foreigners from New England decided that they would help us in our struggle to be free. Somebody needs to help children free themselves from oppression, because they have nothing else, they have no one else. They can only rely on adults to be compassionate.

PP: *When you were writing the book, where do you feel your strength came from?*

AW: My strength comes from the earth, which I really understand. As a child, I bonded with nature very early, because my mother has always been a very great earth spirit. She knew the land and she knew what it produced and she worked with it and she loved it, so this has been my heritage. I believe that the body of woman is our symbol of the earth and the earth's processes. So working on this book, I have been able to think of the body of woman, scarred and battered, as the body of the planet scarred and battered. And to see that helping woman to be whole will help restore her strong connection to the earth and to the protection of the earth, which only women, I think, can feel, in the way that you can empathize with anything that resembles you so closely.

Westerners have just recently, within the last ten years, discovered that the earth is a living being. They talk about Gaia and take this back to the Greeks, who called the earth the goddess Gaia. But indigenous people around the world have always insisted that the world is a living entity and that it should not be abused because it is alive and it feels.

PP: *When we were in London recently at a public meeting where you were talking about the book and about female genital mutilation, I noticed that one of the common responses among women was terror and shock and then a feeling of being unable to do anything. What would you say to women who want to do something about it but feel paralyzed by the fear and the shock of what this is?*

AW: I've noticed that they also faint. Everywhere I have talked about this, women have just dropped. It has been the most amazing thing. It's a real literal shock to the system that this can happen, and you do need time to think about it without feeling pressure. I always tell women when they start reading my book that if the going gets rough, they should just put it aside and not try to press on. Because it is a lot to take in if you are really in touch with your own feelings. If you are not in touch with your feelings, then you can read it as a kind of intellectual exercise. But many women really are not able to do that, and I am happy that they are not, because it means that they are alive.

It is a serious thing that we are talking about being done to the human body. Wherever the human body is, we are one body, and that is why women faint when they hear about female genital mutilation being done to another woman. They know it's also their own body. So, first, just take your time. And because time is pretty much all there is, just take it in, let it go in, and think about it, live with it, sit with it. Then just look around and see where you can be useful. There are organizations like FORWARD International in London that can always use money or help. But mostly be aware and try not to see this just as an isolated brutality that happens to women in other countries but to see how it also happens in this country, in different guises.

PP: *Alice, you've talked about all the different kinds of responses you have had to the book when you've spoken in public. What in particular has stayed with you?*

AW: I was speaking in Washington, D.C., in the early summer to a group of people who were then doing antiapartheid work. I said to them that their work was wonderful and we have to keep doing that, but we also have to see genital mutilation and the abuse of children and women with the same kind of moral ferocity. Andrea Young, the daughter of Andrew Young, was at this meeting. She went home to Atlanta and encouraged her church to become a sanctuary for Somali and Sudanese women in Atlanta. In one instance, a Somali man is refusing to share the lottery he won with his wife. He says she is not really his wife because he was not able to penetrate her, since she was pharaonically circumcised and infibulated. She has taken him to court to prove that she is his wife and that he should share the money with her. What is lovely to me is that, because Andrea Young has made space in her church and in her community for this issue, this woman is not alone. She has support. And this is something that women can do all over the U.S.A., especially in cities like Washington, Atlanta, New York, San Francisco, L.A., which are port cities, where there are an estimated ten thousand children at risk of genital mutilation.

PP: *Alice, how do you feel about the fact that genital mutilation is practiced by women on women and young girls? Mothers doing it to their daughters, and grandmothers doing it to grandchildren?*

AW: I feel terrible about it. It is true that in some places men do it—for instance, in Egypt. There was a recent photograph of an Egyptian doing it to a fourteen-year-old, who had her head against her mother's chest, looking absolutely scared and in pain, and the mother very forcefully holding her while the mutilation is being done by this man. But generally, you are right, this is something that women do. Like everything to do with children, men turn it over to women. Men don't on the whole feed them, bathe them, or brush their hair either. But the mother's betrayal of the child is one of the cruelest aspects of it. Children place all their love and trust in their mothers. When you think of the depth of the betrayal of the child's trust, this is an emotional wounding, which will never go away. The sense of betrayal, the sense of not being able to trust anyone, will stay with the child as she grows up. I think that is a reason why in a lot of the cultures that we are talking about, there is so much distrust, so much dissension, and so much silence. There is all this unspoken pain, this unspoken suffering, that nobody is really dealing with, nobody is airing, and it goes somewhere, it always does.

PP: *During British rule in Kenya, the indigenous people fought to hold on to traditional practices, but unfortunately that included some which were harmful, such as female genital mutilation. This is something you have referred to in your interviews with women here. What are your own thoughts on the subject?*

AW: When they themselves are being oppressed, people tend to hold on to the practices that they can enforce. As they can most easily enforce things that control women and children, that is what they have tended to do. It is very sad, because the British, in this particular instance, were right about stopping this sort of evil thing. But because they themselves were so evil, and the harm that they had done was so great, it was very difficult for African men and women to really choose what they would like to retain of their culture, since the British were so busy destroying everything else. Colonizers have managed, in so many instances, to destroy worthwhile traditions—they have even managed to stop people from eating their own native food, such as their rice. They have made available only less nutritional foods. They have made people stop wearing their own clothing. They have substituted British-made and French-made clothing and goods. Yet the practice of female genital mutilation remains unchanged. The colonizers managed to force people to change many other traditions, but mutilation of girls was clearly not their priority—it seems that colonizers don't care as long as it is somebody else's child who is being abused and they get their profits anyway.

PP:  *There is a very large percentage of women all over Africa who have been either circumcised or genitally mutilated. What do you feel about the fact that it's a culturally specific form of violence against women, yet there are so many other forms of violence against women in other countries? It seems there is a continuum of violence against women that takes specific forms in different cultures. One of the worries I know I have is that it's very easy for Westerners to use genital mutilation as a way of describing Africa as being backward and savage and barbaric, and feeding into all those sort of racist perceptions of Africa.*

AW:  Yes, that is a problem. If you think about the Middle East and Malaysia and Indonesia and countries in Africa where this happens, you could easily think that this is an isolated assault on women. But as we know from our television screen, the assault on women is worldwide; it is not isolated. It varies only by degree.

If you live in a culture like ours, which now wants women who are very thin, very white, very blond, with very big breasts, a lot of women have breast implants, they have bleached skin, they have bleached hair. They try to be the ideal woman for some nonspecific but very powerful man. So it's really about shaping a woman in the image that men think they want. And every country in the world is busily doing that.

As we know, for centuries the Chinese broke the feet of women and forced their toes back underneath the feet so they would only be three inches long, and you had to hobble along on these little feet. Soon this became an aphrodisiac. The fact that your feet were slowly rotting and emitted a peculiarly awful odor became something that men loved. But essentially it was about preventing a woman from getting away. It was about enslaving someone.

And that is what all this is about. It's why we have bras that make it impossible for us to breathe freely, high-heeled shoes that destroy our feet. Some Burmese women wear fifteen or twenty necklaces, designed solely to make it impossible for them to hold up their necks without the support of these necklaces. So if they are ever "disobedient," all someone needs to do is take away a couple of the necklaces, and the excruciating pain of trying to keep their head up makes them long for this necklace again. And there are cultures where they put heavy weights on the woman's legs, which destroy the muscles of the leg. Over centuries, women begin to think that these things are beautiful, because the mind needs to rationalize what is happening—when in fact it is about enslavement.

PP: *Alice, there have been so many different actions, campaigns, and outcries against female genital mutilation, yet it still continues. Why do you think this is, and what do you think can be done to stop it?*

AW: It continues partly because in many cultures there is no role for a mature woman after puberty except marriage. And if you have a culture in which men will not marry you unless you have been mutilated and there is no other work you can do and you are in fact considered a prostitute if you are not mutilated, you face a very big problem. Women mutilate their daughters because they really are looking down the road to a time when the daughter will be able to marry and at least have a roof over her head and food.

Yet you see women and men continuing this practice in countries like Holland and Britain and the United States, where who's to say what a child will do, who's to say that this child is going to be marrying anybody? Who's to say that if she is from Somalia or from the Sudan or from The Gambia or from Senegal, she is going to marry somebody from a culture where this is wanted? In fact, what is happening is that men from their own culture are actually choosing to marry women who have not been mutilated.

In the Sudan and Somalia, a lot of the men marry European women, or they marry African-American women, or they marry Indian women—whoever has not been mutilated. Because often they don't know fully what has been done to women in their own culture, and when they find out, they are horrified.

There are a lot of reasons why it is difficult to stop the mutilation of women, but I also feel that there are a lot of reasons why today we have a greater opportunity to start putting an end to it. It is all about consciousness, and there are now more educated women in the world, more conscious women. In some cultures, men have confused sexuality with torture, so that they only really enjoy sex with a woman if she is literally screaming in pain. But I think that many men are changing, their consciousness is changing, so there will be men, too, who will fight against it, and have been doing so. Sudanese doctors have been against it; in Burkina Faso, Thomas Sankara made a very strong statement against it before he was assassinated. So there are both women and men who are speaking out against it.

PP: *When you were visiting London to promote your book* Possessing the Secret of Joy, *you met with Aminata Diop. Can you tell me about her? What was it like to meet her, and how did you hear about her story?*

AW: I read about her in a newspaper. She is a young woman who ran away from Mali to avoid being mutilated. She went to France, seeking political and religious asylum. They decided she couldn't be granted either. I was interested in supporting her, and so I met her in London.

My feeling was what I imagine freed people felt encountering someone who has escaped slavery. This feeling was probably intensified because we met in London, which had been the home of the antislavery society. Frederick Douglass, for instance, talks about getting to London and meeting people from the antislavery society and telling his story. So there was a feeling of meeting someone who had escaped enslavement, someone who had the courage to decide that she wanted to keep her own body intact. And someone very harried and very broken, too, because, as she said through tears: "I miss my mother, and my mother believes that because I refuse to have this done to me, I am a prostitute." This is really almost unbearable. We said to her, We will go to your mother, we will make her understand that you have done something that is noble, something that is really good. And she said: "No, my mother would never understand, because she believes that mutilation is written in the Koran. She can't read it herself, she is illiterate, so she has to take the word of the imam and my father."

So there was a painful side, but the other side was just incredible joy. There is a limited amount that anybody can do in the world. But the one thing everybody can do is to alleviate the suffering of some*one* else or some*thing* else. You may not be able to take away all of it, but you can make it much, much less. It's something you can do every day. And the pleasure of being able to do that is so great, it releases an energy that can change more than you may imagine.

PP: *What are your expectations of going to Africa and talking to women in Africa, both those who are campaigning against genital mutilation and those who have had it done to them?*

AW: I am a great believer in solidarity. Nicaraguans say something very beautiful. They say that solidarity is the tenderness of the people and real revolution is about tenderness. The sharing of this tenderness is beautiful. If you can make one person's life free from a particular kind of pain, that's really enough. It may have ripples, it may go here, it may go there; there's no way of knowing. I think it is the nature of life to continue what is put into action. You may not be around to see where your own action goes—people never are, really, because it never ends.

I'm looking forward simply to being with the women we will be talking to. I want to eat with them, dance with them, see who the priestesses are and who the goddesses are. I want to know what's going on with them and how they feel. If they tell me that they would prefer that I didn't intrude into their affairs, I won't talk to them about that. I want to ask them, do they know what African custom made it possible for me to end up on another continent, in the U.S.A.? We have been separated by a custom similar to genital mutilation, a custom of slavery. I have been wondering about my ancestor who came from Africa to America: was she mutilated? What did she feel, in chains and mutilated in this way? If she had been pregnant, she might never have made it alive; she might have died trying to give birth on the ship, with nobody to open her if she had been infibulated and sewn together.

PP: *Can you tell me why you wanted to do a film about female genital mutilation and what hope you have of this film making a difference?*

AW: I always think about people who don't read, who cannot read, but who can relate to things visually. This is because I myself come from a culture where not everybody reads, and a family where not everybody reads, so film is very important. It is very important for people to be able to actually look at what it is you are discussing and to understand it because they see it. I hope that most of the women in the West from mutilating cultures will eventually see this film, and I think they will. I think they will be curious. They may not care much for my perspective or what the film is supposed to be about, but I think that out of curiosity they will want to see, because it is about their own experience. And they will be able to enlarge upon it because they will be able to take issue with it, they can criticize it, they can use it, as an object from which to explore their own experience. And that is what art can do. So I have that hope.

I also think that because it will be portable, it will be able to go into countries where there are a few women who are against genital mutilation and who will be courageous enough to show it to women who don't know really what they feel about it. Many of them have forgotten the pain they felt when it was done to them, because they were so small. And now, of course, they do it to much younger children, to babies. And if it is done to you as a baby, you can only imagine what you retain of this assault. If, in fact, you live to know about it. Because in France, a

Gambian woman is on trial for mutilating a baby, who bled to death.

It's also a good thing to have a film that shows that the genital mutilation of women is really just a part of the global mutilation of women, the terrorization of women, one of the numerous things done to keep them in their place, under the foot of the dominant patriarchal culture.

# Alice Walker and Awa Thiam

*Awa Thiam is well known for her strong feminist writings in support of women's rights in Africa and for her commitment to the eradication of female genital mutilation. I understood she had suffered a great deal for attempting to bring women's issues into the foreground of African thought. I had long desired to meet her and to embrace her as a sister in struggle. Today she is, among other things, a politician. (Interview questions prepared by Pratibha Parmar.)*

*Pictured below, from left: Nazila Hedayat, Awa Thiam, and Alice Walker, after interviewing Awa at the Savana Hotel in Dakar, Senegal.*

AW:  *Awa, what is your personal involvement in campaigning against female genital mutilation?*

AT:  I am a member of a tribe which practices up to 80 percent female circumcision and sometimes infibulation. Because of that, I became involved in campaigning against the perpetuation of these practices.

AW:  *There are many different approaches to fighting the practices. What in your opinion is the most effective way of challenging and changing this tradition?*

AT:  It seems to me that if you start from the feminist experience, you can in the first instance struggle simply among women, telling yourself that women constitute a force able to question the patriarchal system. I was militant for quite a long time strictly among women to fight against the practice of female circumcision and infibulation. But after more than ten years' struggle, I realized that it is necessary to pass over the female stage—in other words, you have to leave the female circle, to get involved in the sphere of politics, because there you find the decisionmakers. I told myself that it is there I have to be—to succeed in convincing the decisionmakers, both male and female, and try to struggle for the abolition of sexual mutilation.

AW:  *Why is it still practiced?*

AT:  I think the answer is contained in the questions "Why is it that everywhere women are dominated? Why do they continue to be submissive? Why is it that on all five continents one finds a quasi-identical situation, different only in its form? Why is it that women always end up doing the domestic jobs?" You can go wherever you want—to America, France, India—and everywhere you will find women in the middle of doing domestic chores: that's a common aspect everywhere. Another thing is that women on all five continents are always subordinate to men. The subordination we're speaking about exists everywhere.

AW: *Have you seen any change in attitudes, and do you think that the practice is decreasing?*

AT: If I consider the intellectual milieus which represent a very small portion of societies that practice this sexual mutilation, I realize that there is something small happening which is positive, although one has the impression that nothing is being done. But to succeed in fighting against sexual mutilation, you have to get the women together, make sure that they meet, even if they just discuss the weather and how to survive—but at least make sure they meet and find work they can do together, because the more they work, the stronger those links between them become. And from then on you can start working on their awareness.

If you are starving, you cannot listen: the only problem is survival—to find something to eat straightaway. Right from the beginning, my strategy has been to get the women together and then to see how to work on the level of ideas, how to succeed in abolishing all these practices which are so bad. That's one perspective. The second perspective is to expand, to broaden out from the specific context of women and to look at all the different strata of society, where there is power, any power which can lead to a radical change in the situation of women, not just female circumcision and infibulation but also oppression of women in all its forms: polygamy; discrimination at work; the sexual division of labor; battered women; all forms of sexist violence; sexual harassment. We would need a lot of time to eliminate all the forms of patriarchy!

AW: *You have said that you have worked with women for over ten years and you have wanted to broaden the base that women can draw from, to include religious organizations and other political institutions. What kind of stronghold do the Muslim clerics, the* marabouts, *have, and where does the power base lie?*

AT: I have to say that the *marabouts* in Senegalese society represent a real power. As an example: if today all the *marabouts* said you have to vote for a particular political leader, he would be elected. So they are people who represent a real force. Patriarchal power is incarnated in Senegalese society through the state and the institutions, but I would say first and foremost through the religious institutions. If today all the religious leaders said in the media that it's necessary to suppress female circumcision and infibulation, you could be sure that in less than two or three years, people would have finished with the practice of female circumcision and infibulation.

But the question is: Why am I trying to work with the religious leaders and politicians? Quite simply, because I have discovered perhaps quite late that they represent a real power, and as far as politicians are concerned, it's exactly the same thing—I thought that women were sufficiently strong to question the patriarchal system. I still think so, but we have to refine our strategies, we have to learn where we have been in order to know where we are heading.

AW: *I am very curious about what the response is to you from these religious leaders and what impact you feel you personally have. Do they seem to be people who can be persuaded to change?*

AT: I am a researcher, and as a part of my work I can go to anyone, no matter who, and can disturb whomever I choose. Recently I had to meet the leaders of the large religious organizations in Senegal. I was very well received, I was given authorization, and therefore all doors were open to me. I took the time to speak and was accompanied by a technician, because he introduces me and asks questions. Given the patriarchal context, generally speaking the man, the technician, introduces you, and then the woman can speak. It is easier to start a dialogue in this manner.

I asked precise questions about the position of Islam in relation to female circumcision and infibulation. What does Islam think of these practices? What does the Koran say? I admit that I was dealing with honest people. The religious leaders of this country are honest answering those questions.

AW: *What are all the reasons given for the continuation of this practice?*

AT: As well as religious doctrine, there's a second point which appears to me essential. The oppression of women by men is expressed through this practice. When you go back as far as possible in history, you realize that even before religion, before Islam, female circumcision and infibulation were practiced. The texts are there that simply prove that man has always wanted to control woman, control her body as well as her mind. When you cut off a woman's genitals, when you sew them together, when you open them to have sexual relations, when you sew them up again when the husband is absent, open the genitals again to allow her to be penetrated by her husband, there's no need for explanation—everything is clear. You control the woman as you control no matter what object, no matter what possession or property.

AW: *Many women believe that they are defending a fundamental traditional value when they defend genital mutilation. How can these women be challenged, and how can the religious fundamentalists be challenged?*

AT: It seems to me that it's a problem of mind-set, for the fanatics as well as for the women who continue to perpetuate this practice in good faith, telling themselves that they must do it. Work can be done with groups of women on this mind-set. We must give valid arguments, solid and convincing, to succeed in putting an end to this vicious circle. I think there are no worse enemies for women who are struggling for their rights than women who totally agree with the patriarchal ideology, who willingly take the men's side against women who want to have their rights respected and strengthened.

AW: *Your book, which was published in 1978, was one of the first books to articulate African women's concerns about female genital mutilation. What was the response? Have you suffered any consequences as a result of your radical views and your courage in presenting this information?*

AT: The responses have been various. The work was published in France and very well received—not only in France but elsewhere in Europe—and has been translated into English and German. On the other hand, the reactions in Africa, I have to say, have been rather mixed—rare were the people who appreciated the contents of the book. There were certain people who wanted to restrict the content to the question of circumcision and infibulation. But in this publication I tackle the question of the oppression of women in general, of patriarchal oppression.

I have to say that I find in some African intellectuals, men and women, an indescribable dishonesty. They have the tendency to say that struggling for women's rights is something specifically for Western civilization and has nothing to do with Africa, the real Africa, traditional and traditionalist Africa. I find that very wrong. Liberty concerns all of us, and fighting for universal rights is also fighting a universal struggle. One cannot be an intellectually honest man and not recognize the fair struggle of women to have their rights respected.

AW: *Do you see this as an international issue that women all over the world could organize against?*

AT:    Circumcision and infibulation constitute for me the most eloquent expression of oppression of women by men. So this concerns us all, and those who are aware of their rights should mobilize against this. All of them. Europe did not wait for African women to cry for help; European feminists mobilized fairly quickly right from the seventies. It started in the United States, with Fran Hosken, who from 1972, with her magazine *Wom-News,* started to handle this question of sexual mutilation. France invested a lot in the struggle for the abolition of circumcision and infibulation. The various ministries with women, notably that of Madame Yvette Roudy, invested effort in the struggle for the abolition of sexual mutilation by financing international meetings. France really has done a lot against sexual mutilation, and it seems to me that other European countries are doing likewise, but hats off to France.

Any definitive and irremediable removal of a healthy organ is a mutilation. The female external genital organ normally is constituted by the vulva, which comprises the labia majora, the labia minora or nymphae, and the clitoris, covered by its prepuce, in front of the vestibule to the urinary meatus and the vaginal orifice. Their constitution in female humans is genetically programmed and is identically reproduced in all embryos and in all races. The vulva is an integral part of the natural inheritance of humanity. When normal, there is absolutely no reason, medical, moral, or aesthetic, to suppress all or any part of these exterior genital organs.

—Gérard Zwang
*Mutilations Sexuelles Féminines,*
*Techniques et Résultants*

# Pratibha Parmar and Dr. Henriette Kouyate

*Dr. Henriette Kouyate is a gynecologist based in Dakar, Senegal, who has been working against female genital mutilation since 1955. She runs the Sokhna Fatma Clinic, where she treats mutilated women, and organizes retraining workshops with circumcisers. She is also the general secretary of COSEPRAT, the Senegalese National Committee on Traditional Practices.*

PP:  *Can you tell me what you do and how you became involved in this issue?*

HK:  I'm a gynecologist specializing in obstetrics. And when you take care of women, what interests them interests you. Through that I became involved in looking into the traditional practices which have an adverse effect on the health of women—among others, circumcision and infibulation.

PP: *And the organization for whom you work: what is its name and what are its objectives?*

HK: I am the general secretary of the Senegalese National Committee on Traditional Practices. This committee was born following the conference in Dakar on traditional practices in 1984. We decided to form a follow-up committee, because many conferences had taken place before but without any follow-up. Since then, in most countries which were represented there, such as Somalia and the Sudan, national-level committees have been created to focus on traditional practices which have an adverse effect on women.

Our most important objective is to inform people—to make people aware and to educate them. We want to reach ordinary people, meet the religious leaders, the political personalities, the young, adults, women, children—to inform them, to talk with them and show them the bad side of these practices—while trying to promote traditional practices that are beneficial for the health of women and children.

PP: *How do you do this? Do you go to the villages? What kind of activity do you organize to achieve that?*

HK: We have seminars for people in positions of responsibility: midwives, doctors, cultural advisers. We ask them to spread information and to make the population aware. Through that we create committees at the regional level. We also had a seminar for people performing circumcisions. They came from different regions, especially here in Senegal, where there are many women and it is in their tradition to be circumcised—from the mother to the daughter to the granddaughter.

They came and exchanged ideas. It was very important, because they have a position in their society. For them, circumcision is a duty they have to perform. That is the tradition they have been taught, and so they carry it on. On the one hand, they think it is tradition; on the other hand, they say it's religion. They do it from mother to daughter, and they don't know why they would *not* do it.

Our task was to explain to them the harmful side of their attitude, to explain that they are not obliged to do this and that it is not a religious obligation, and to show that it is not possible to perform a circumcision without causing harm. We showed them the route of the principal vessels, why there is hemorrhaging, why there are infections, why this is an assault against women's sexuality, and why it is a problem which needs to be discussed.

They accepted and understood, although in the beginning there were some who did not want to hear. For them, it's a vital problem, because they earn their living by performing circumcisions, but those women also have another activity. They are at the same time traditional midwives, trying to improve the delivery of babies. They are beginning to move away from practicing female circumcision and instead are focusing on using traditional methods for delivering babies.

The argument they put forward was that by circumcising women they were giving them a certain education. But this is not true, because circumcision is performed between the ages of one month and five years. What kind of education can you give to a small child?

They will tell you that it's done to preserve virginity, but how many young girls who have been circumcised have become pregnant outside of marriage? I believe the women were happy to have a chance to talk, because many have realized that in reality, by practicing circumcision they only created problems. When they discussed this with other women, they realized that something was not right.

PP:  *Dr. Kouyate, you have lived here for several years. Could you tell us how long you have been working on this question and tell us some stories regarding women's circumcision, the complications and problems you might have encountered.*

HK:  I have been working on this since 1955. I originally lived and worked in Mali, where I have looked after women in many ways: giving birth, sterility, the interval between having children, and, of course, the problem of women and circumcision. At that time, it was not possible to discuss circumcision or polygamy. But we tried a little to advise people, and some people were persuaded not to have their children circumcised. There are now women in Mali who were born before the sixties who were not circumcised.

It was not possible to discuss circumcision in public; just mentioning the word was daring, as it related to women's intimacy. All women in Mali were circumcised but did not think they had suffered, because women would never talk about themselves, about their sexuality, the problems encountered in their intimate lives. But now one can speak more openly about circumcision, family planning, the interval between having children, and abortion. It is important to talk about these things, because we have to face reality. Many children have died of hemorrhaging after being circumcised. I know a woman whom I treated who had fertility problems. When she did eventually get pregnant, she gave birth to a little girl. But one day when she was away they came to take her little girl—they circumcised her, and she died.

There are so many stories like that. The parents could not make an official complaint, as it was a family problem. In Africa, no one takes in-laws to court, saying, My child has been taken away, has been circumcised, the child has died. But it's not right, either, that the in-laws have so much power over children and even if they die it is kept quiet.

PP: *Doctor, could you talk a little about the medical consequences of circumcision and infibulation?*

HK: First, you can suffer fatal hemorrhaging, because in the clitoris there are a lot of blood vessels, including the dorsal artery, the vein of the clitoris, so young girls can bleed to death. Fear is also a very important problem, because when those children are taken away they are not prepared for the pain they are going to suffer, and the pain creates stress and shock. And in addition you have infections, even tetanus. Infections begin in the area of the wound but may spread to the internal organs.

Besides circumcision there is the other, much more dangerous practice—infibulation. While circumcision concerns the clitoris and the removal of the clitoris, with infibulation not only the clitoris is removed but the small lips—the labia—are cut off and the big, outer labia as well.

And the last straw is that what is left is closed up, using thorns or whatever they can find. They just leave a small orifice, through which the menstrual blood can flow. But frequently this orifice cannot let out everything. So a mixture of blood remains inside the vagina. As a consequence, a painful infection develops, which can cause sterility.

But that is not all. The woman has been cut and traumatized, so intercourse is very painful and there are a lot of problems giving birth. An area which is normally elastic has become a cicatrix area. As a result, many women tear, at the top and at the bottom. If they are in hospital, you can perform an episiotomy, you can enlarge the opening. But just imagine these women who have to give birth at home, where there is no notion of surgery. They are just left with a tear at the top and bottom, at the top causing massive hemorrhaging, at the bottom even worse—the tear can enlarge, extending to the anus, so that women can no longer contain their feces. When the tear is central, it expands. Not only are women affected, but babies are affected, too. Because the birth takes much longer, and that can cause problems for the baby.

PP: *You spoke the other day about the question of girls who get married very young.*

HK:   The problem of circumcision is not the only problem which interests us. We are also concerned about the number of girls getting married as young as ten years old and then getting pregnant while they are really still children themselves. And those pregnancies very often finish very badly. For the mother as well as the baby, because they are girls whose bodies are not fully developed by the time they are forced to bear children.

As a result, many girls have abortions, or they may have nervous breakdowns, and the mortality of the mother and infant is very high. Young girls may have vascular vaginal fistula because the womb is too small, the child cannot come out, so when they are giving birth, the vagina may burst into the bladder. It is very traumatic, because those women are going to leak all their lives if they cannot go to surgical centers where doctors can try to renew the bladder. So we are also involved in fighting against child marriages, because they are harmful to the health of our women and children.

Another problem preoccupies us—the pressure on African women to have a lot of children in a short space of time. Women need to recuperate between children, for the sake of both the mother and the child. If births are spaced out over three to four years, two years minimum, the mother will have time to recuperate and the child will be born healthy.

PP:   *You made some connection between AIDS and infibulation. Can you tell us something about it?*

HK:   I am convinced there is a connection between AIDS and female circumcision. Look at the conditions in which circumcision is carried out. It is a communal circumcision: the circumciser has her own blade, she cuts and passes from one child to the next with the same blade, soiled with blood, the hands soiled with blood. So it's evident that if she is a carrier of AIDS, if she cuts herself she can transmit the disease. Or if one of the children being circumcised is a carrier she can transmit it.

We think there is a real change—I will give you one example. We were asked to give a talk in a center in Dakar. In this center there were a lot of women from a tribe where about 75 percent of women are circumcised. A middle-aged man of that tribe wanted to kick us out. We refused to leave because it was a social center. He asked the women there not to watch the film they were showing, *The Deceit of Circumcision.* He gave them the order to leave, but they refused to leave. That is very important: they wanted to listen and to know about circumcision. They had been circumcised themselves but wanted to think about the future with their children. There were even two women who had the courage to say they had daughters they did not want to have circumcised.

PP: *You said the other day that the best place to discuss these problems with women was during childbirth. Can you tell us something about it?*

HK: Yes. In the delivery room is the right time, even if it sounds a bit vicious to make women understand that circumcision, by reducing the vaginal passage, creates an unnecessary complication. We are forced to cut circumcised women, to perform an episiotomy. They realize that if they had not been circumcised, we would not have been forced to operate to allow the child a normal passage. We explain to them where the child is stuck—and what for? So they understand and I ask them not to have their daughters circumcised. Many tell me that they will not have it done—and I think they are sincere. We have achieved a lot of success by word of mouth, by explaining. It is a question of patience, but you have to be careful not to upset them, not to humiliate them.

PP:  *Why do you think circumcision persists, despite efforts to stop it?*

HK:  Many factors are put forward, such as religion, tradition, or hygiene. But the real problem is the need to control the sexuality of women, to control their desires, to try to keep them like children, like someone with no responsibility of her own, who cannot be a human being in her own right. An adult woman is perfectly able to control her sexuality, but circumcision is maintained to control the sexuality of women. And I think that many women no longer accept being treated like children and fight against it. It is a fundamental matter of principle that we refuse to let our sexuality be controlled. Because we are responsible human beings, better able than men to control our own sexuality.

PP:  *Can you then say that this question comes from the patriarchy— that it is a kind of specific control of women exercised by men?*

HK:  Yes, that's exactly what I am telling you. It's a matter of control. It is the desire and need of men to control the sexuality of the woman. We now refuse this permanent control. And many people, even men, are beginning to understand this.

PP:     *Is there anything else you would like to add?*

HK:     Our message is that we ask our brothers, the men—we do not compete with them, we are complementary—that they try to understand women. Because if you want to work together, there must be understanding and mutual respect. If you are kept as irresponsible children you act as irresponsible children, but if you are given responsibility and treated as an adult, then you act as an adult. African women are adults—it is time this was recognized.

# Alice Walker and Circumciser 1

*Perhaps the most difficult of all the interviews I did, this one, in Dar Salamay, The Gambia, was in many ways <u>the</u> interview I came to Africa to get. Why have women become the instruments of men in torturing their daughters? What would such a woman have to say for herself?*

*Pictured below: Interviewing the circumciser in her front yard. Her assistant is at left. In her hand she holds her "crown" of authority.*

Q:    *I would like to know how you became a circumciser.*

A:    *(translator)* She says her mother was a circumciser, her grandparents, her great-grandfather, were circumcisers. So when her mother died, she was elected as the circumciser.

Q:    *I see.*

A:    Of the village.

Q:    *And were you pleased to become the new circumciser, or was it something that—*

A:    Yes, she was very happy.

Q:    *And how long have you been a circumciser?*

A:    She's in her fifth year now.

Q:    *And how many girls have you circumcised, do you think?*

A:    She says she cannot remember, but she's taken so many children now, she doesn't—she cannot remember exactly how many.

Q:    *I see. Hundreds or thousands?*

A:    When they are taken, the children, to be circumcised, they don't have an exact amount. Sometimes they take about fifty children. But after those children are taken, twenty more would come, and ten more would join them, so they wouldn't know exactly how much.

Q:    *And how often did you do it?*

A:    It is normally during the holidays. Now that these children have gone, maybe by Easter they will take more children.

Q:    *Why is it done? Why is circumcision done to young girls?*

A:     It's their tradition. The grandparents, their folks' parents, have been doing it and inherited it from them.

Q:     *Do you think it is a good tradition, and why?*

A:     She says they're doing it because it's their tradition. She says normally during the healing period, and during the coming-out ceremony, there's a lot of food, a lot of festivity, and they like it.

Q:     *If you could say how she does it, what she does, and whether she would be willing to show us what she does.*

A:     She says all her instruments are in that calabash. She says even the girls that are circumcised, they don't know what is in it. And they don't know what she uses. So it's—it's a secret, nobody knows what is in that thing, and she's the only person who knows what is in that calabash—she and her sisters.

Q:     *And we . . . ?*

A:     But she would not allow anybody else to see.

Q:     *Could you tell us what is removed from the children, what part of the body is removed from the children?*

A:     She says that also she will not reveal, because it's—it's a secret, and it's a secret society. It's not to be revealed to anybody.

Q:     *Well, we know that when children are circumcised, the clitoris is removed, the inner labia, the outer labia. Sometimes there is infibulation, which is a stitching up. This is not secret—everyone knows this already. What we don't know is why it continues to be done, because it is a very painful thing and sometimes it presents great health problems to the children.*

A:     She says—they say—normally they say it's a small world; well, it's still a big world. She says not everybody would reveal their secrets. Some people will tell you what they cut and how they're doing it and the types they're doing. She says even if the children that are circumcised now, even if you put a knife on their throat, they would never tell you what was done to them. It's—it's painful, but they would never say it. It's—it's their tradition.

Q:     *Does she get paid for what she does?*

A:     She says when they're taking them to be circumcised, they don't pay her anything, she's not given anything. But during the—the washing ceremony, that's the time each mother brings in five dollars. And that's all she gets from it.

Q:     *When you are performing this operation and the children begin to scream and to cry, how do you feel?*

A:     She doesn't feel anything. She says she doesn't feel anything because she—she had experienced it and her mother did it to her, so she feels it—it's not a harm, not harm to them.

Q:    *How many of your children that you have done this to have died?*

A:    She says none of the children have ever died.

Q:    *How many assistants do you have helping you?*

A:    She says when she's really in it, she doesn't know how many people are assisting.

Q:    *What do the assistants do? Do they help to hold the children?*

A:    She says sometimes she does it herself, alone. Nobody helps her to do it.

Q:    *How does she keep the child from crying and trying to get away?*

A:    When she's doing it, without any assistant, anybody, she said that the child would be there, and she would shake, and she knows what she does. The child, she knows. And even if the child is crying, she doesn't know whether the child is crying or not. All she knows is she's doing the operation and the child is not moving and she doesn't hear the child cry.

Q:    *What does she use to stop the bleeding? Does she have an herbal medicine?*

(No answer.)

Q:    *Could you tell me what kind of work you were doing before you became a circumciser?*

A:   She says she was a farmer before she became a circumciser. But now she's old, she doesn't go to the farm anymore.

Q:   *If you could choose a different kind of work than circumcising, what would it be?*

A:   She says if she had a choice of getting another job, she would still be a circumciser.

Q:   *You realize that there are many women in the world who are not circumcised. What do you think of uncircumcised women?*

A:   She says she doesn't know anything about those women who are not circumcised, and she would not go anyplace where those women are. She would not go to them. She would not go to any gathering where those women are. She would—she doesn't know anything about it.

Q:   *Many of us are not circumcised.*

A:   She says she would know—all of us sitting here—she would know which of us hasn't been circumcised.

Q:   *How would she know?*

A:   She was taught by her great-grandparents the art of knowing who is not circumcised.

Q:   *Okay. Am I?*

A:   She says, You want to know everything from me? (Laughs.)

Q: *Do you maintain a relationship with the girls after you have circumcised them? Do you see them, do you take a special interest in them?*

A: She says tomorrow they—all the children—are going back to their parents. She says after that, anywhere they see her, they—they're going to respect her, and the whole village would respect her. They would grow up to respect.

Q: *But doesn't she think that they actually fear her?*

A: No, no, they will not fear at all.

Q: *What is that that she's got in the bottle?*

A: She says it's—it's purified water.

Q: *Many of us notice that a person has been circumcised, because the light in the eyes goes out. There is an absence of light in the eyes. Have you noticed this in children that you have circumcised?*

A: Maybe you—you notice that, but she hasn't. As far as she is concerned, they are happy, and they—they're all excited.

Q: *Is she a happy woman? Is she a happy person?*

A: Yes, she's a very happy person. She says she is happy because she has taken so many children, all of them are feeling OK, they're all well—no, nothing is wrong with them. They're—they're all healthy.

Q:    *Is there anything else that she would like to say?*

A:    She has nothing more to say.

~~~~~~~~~~~~~~~~~~~~~~ Description of an Infibulation

The little girl, entirely nude, is immobilized in the sitting position on a low stool by at least three women: one of them with her arms tightly around the little girl's chest; two others hold the child's thighs apart by force, in order to open wide the vulva. The child's arms are tied behind her back or immobilized by two other women guests.

The traditional operator says a short prayer: "Allah is great and Muhammad is his Prophet. May Allah keep away all evils." Then she spreads on the floor some offerings to Allah: split maize or, in urban areas, eggs. Then the old woman takes her razor and excises the clitoris. The infibulation follows: The operator cuts with her razor from top to bottom of the small lip and then scrapes the flesh from the inside of the large lip. This nymphectomy and scraping are repeated on the other side of the vulva.

The little girl howls and writhes in pain, although strongly held down. The operator wipes the blood from the wound, and the mother and the guests "verify" her work, sometimes putting their fingers in. The amount of scraping of the large lips depends upon the "technical" ability of the operator. The opening left for urine and menstrual blood is minuscule.

Then the operator applies a paste and ensures the adhesion of the large lips by means of an acacia thorn, which pierces one lip and passes through into the other. She sticks in three or four in this manner down the vulva. These thorns are then held in place, either by means of sewing thread or with horsehair. Paste is again put on the wound.

But all this is not sufficient to ensure the coalescence of the large lips; so the little girl is then tied up from her pelvis to her feet: strips of material rolled up into a rope immobilize her legs entirely. Exhausted, the little girl is then dressed and put on a bed. The operation lasts from fifteen to twenty minutes, according to the ability of the old woman and the resistance put up by the child.

—Alan David
*Infibulation en République
de Djibouti*

# Alice Walker and Two Recently Circumcised Girls

*After the interview with the circumciser, I decided to interview two of the girls who had undergone mutilation two weeks before.*

Q:   *Could you tell me what you have been doing for the past two weeks?*

A:   We have been circumcised, and during the past two weeks we have been playing and learning how to dance.

Q:   *Did you know beforehand that you were going to be circumcised, or was it a surprise?*

A:   We were not told that we were going to be circumcised. They just told us that we were going to play.

Q: *And so when you got to the place where you were circumcised, was it a very big surprise?*

A: We were not surprised.

Q: *Not surprised. Were you frightened?*

A: We were not scared.

Q: *What kinds of things did you learn while you were out in the bush?*

A: We were being taught how to respect elders, how to talk to elders, and how to live in the society.

Q: *All in two weeks. Did you leave school for just two weeks, or was it a holiday?*

A: We were taken out of school.

Q: *And did you have many teachers or one teacher here?*

A: There were many people.

Q: *Did you enjoy it? Did you enjoy the studying and the learning?*

A: Yes, we were very happy.

Q: *And did you miss your mothers?*

A: No. Our mothers came every day to visit us and bring us food.

Q: *Did you help look after each other when you were . . . you were . . . ?*

A: Sometimes we help each other, but normally we have some people, some other girls, older girls who supervise us and teach us what to do. They're the "discipliners."

Q: *Was it explained to you why you were being circumcised?*

A: We were not told.

Q: *No. And nobody ever said anything about a reason for it? Other than that it is a tradition?*

A: No.

Q: *So you do not know why this was done to you?*

A: We were not told. I didn't know why we had been circumcised. We were just circumcised.

Q: *Well, would you like to know why?*

A: We would not want to know.

Q: *You would not like to know? You don't think you could ask your mother to tell you?*

A:    Even if we asked our parents, they would not tell us.

Q:    *Ah, they would not tell you. Is this a happy day for you? How do you feel about the stay and the celebrations?*

A:    We . . . it's a very happy day for us.

Q:    *Is this something that you would do to your own little girl?*

A:    Yes.

Q:    *Why?*

A:    It has been done to our mothers, and our mothers did it to us, and we will do it to our children.

Q:    *You realize that many people in the world do not do this to their children. What do you think of that?*

A:    I don't know anything about those children and those women who haven't been circumcised. It's our tradition; maybe it's not *their* tradition—that's why they're not practicing it. But our tradition we will practice and we will see that it continues.

# Pratibha Parmar and Circumciser 2

*This circumciser was attending a workshop in Banjul, run by women campaigning against female genital mutilation. I wanted to interview her because among the circumcisers present, she was the most belligerent and the least willing to listen. I was told that she was responsible for the death of a young girl who had been brought from Paris to The Gambia for mutilation only a few months prior to this interview.*

Q: *What are your reactions after today's meeting?*

A: I am very happy, it is a nice meeting, a very nice occasion which has gathered many people.

Q: *How long have you been practicing circumcision?*

A: I have been practicing for twenty years.

Q: *Following the discussions during the meeting today, do you have new ideas and would you abandon this practice?*

A: No. I have only one idea. I want to carry on practicing circumcision.

Q: *Is circumcision the way you earn your money?*

A: Circumcision not only brings me some money but circumcision is also considered amongst the public as something beneficial to our society.

Q:    *Have you ever experienced accidents when practicing, paralysis or death?*

A:    No. Never. I have been practicing over twenty years, and I have only had success.

Q:    *How did you learn practicing circumcision? Did you inherit it from your family?*

A:    Circumcision is a traditional practice inherited from our great-grandparents.

Q:    *Will your children be practicing circumcision?*

A:    Yes. I will teach them how to practice.

Q:    *When did you practice circumcision last and on how many people?*

A:    The day before yesterday, Saturday, I circumcised two children.

Q:    *What kind of material do you use for circumcision?*

A:    The blade.

Q:    *What kind of precautions do you take, when using the blade, to avoid infection?*

A:    I boil them in water, I dry them, and I put them back in the package.

Q:   *How many blades do you use?*

A:   I always buy a package of blades and use one blade for two people.

Q:   *How do you feel when children cry during the practice?*

A:   The people I circumcise do not feel tired, because I have very big experience in practicing. Those children can eat straightaway and go to play with their friends.

Q:   *Is circumcision painful?*

A:   It is unavoidable. . . . The flesh is cut without anesthetics, it must hurt, but the pain does not last.

Q:   *Do you know that the practice of circumcision has been abandoned and condemned by many people throughout the world? What are your reactions?*

A:   I have nothing to say about it.

# Pratibha Parmar and Circumciser 3

*I interviewed this circumciser in Dakar, in the courtyard where a women's meeting with Awa Thiam was held. She was in her nineties and had to be helped to walk by her daughter, who is taking over as circumciser. She brought a bundle, which she carefully unwrapped. She showed us her knife, her dirty cotton wool, a wooden stick that she put between the girls' teeth to stop them from screaming, and a half-used tube of cream. She was the only circumciser who was willing to show us her instruments and to talk about how she cut the girls. This was a very chilling and harrowing interview, as she kept picking up her knife. (It is <u>her</u> knife that appears throughout the film.)*

Q:     *What do you do in life?*

A:     I am a blacksmith.

Q:     *How long have you been a circumciser?*

A:     I practiced it on two young girls yesterday.

Q:     *Who taught you how to do it?*

A:     As we are blacksmiths, my mother was the one with the secret. After her death, the circumcision knife was given to me so that I could carry on the practice.

Q:     *When did you start practicing?*

A:     One year after I was married, I was given the knife.

Q:    *How many girls have you circumcised?*

A:    So many I can't remember.

Q:    *How do you do it?*

A:    When I am ready with the knife, I use the water from this bottle to use on the eyes. I do it with this knife, and afterwards I put some cotton wool and alcohol on it. I perform it, and afterwards I spit and the hemorrhaging stops.

Q:    *If you see a girl crying, what do you feel?*

A:    I can tell you that all the girls that come to me do not feel any pain. Because I take all the necessary precautions. You can ask them without my being present, and they would say exactly the same.

Q:    *Do you let them go home after the operation?*

A:    Yes, after the operation I make a talisman, and then they walk home.

Q:    *Do they have to be defibulated before they get married?*

A:    I do not know.

Q:    *When you circumcise, what do you cut?*

A:    There is a small part that is shaped like a horn, which I cut, then I wash with salted water before applying alcohol.

Q:   *Do you get paid?*

A:   Yes; that's how I earn my living. I pay for my home and my clothes. Everything comes from this knife.

Q:   *Have any girls died after you have done this to them?*

A:   Since I have started practicing, I have never had an accident. No one has ever died in my arms. When I circumcise, I take my knife when I am ready. I am very precise. I use cotton wool, alcohol, and penicillin. I ask the child to lie down, and I remove what has to be removed. I read verses from the Koran, and I have never had any problems.

# Alice Walker and Mary

*Mary is the mother of a four-year-old daughter, also called Mary, who had been circumcised two weeks before this interview.*

AW: *Mary, why is this such a special day for you?*

M: Because our daughters have been circumcised and today's the coming-out ceremony.

AW: *And why is this so wonderful, what's so wonderful about it?*

M: It's a wonderful day because it's a tradition and it's been a tradition since the days of our grandparents.

AW: *And was your daughter afraid when she went, or didn't she know what was going to happen?*

M:     She wasn't told.

AW:    *When you went, did anybody tell you?*

M:     No.

AW:    *And what did you think was going to happen?*

M:     I didn't know.

AW:    *What did they tell you—how did they get you to go?*

M:     I was told, "Where you're going to be taken, you're going to eat a lot of bananas, a lot of food."

AW:    *And when you got there, what happened?*

M:     I just felt something that hurt.

AW:    *Were you very surprised?*

M:     Yes.

AW:    *Were people holding you down so that you couldn't get up?*

M:     Yes.

AW:    *Did this frighten you?*

M:     Yes, I was very frightened.

AW: *And were they people that you knew?*

M: No, I didn't know them, and my eyes were tightly shut.

AW: *Was it hard for you to let your child go?*

M: Yes, I felt sorry for her.

AW: *But you couldn't keep her at home?*

M: I felt sorry for her, but I couldn't keep her, because it's the tradition and the child had to go.

AW: *What would happen if you did not let the child go?*

M: Here there are many events and places where a child can't go if she isn't circumcised.

AW: *What are the things that she cannot participate in?*

M: Ceremonies like these coming-out ceremonies.

AW: *And what about marriage later—would the child grow up to be able to be married?*

M: Even if she isn't circumcised she could get married.

AW: *How do you think little Mary will feel when she sees you?*

M: I'll take her some food and sweets so that the minute she sees me she will forget about the pain she has suffered.

AW: *Do you think that Mary will feel that you misled her and that you betrayed her?*

M: Perhaps, but later on she will know that it's tradition which demanded it.

AW: *Who do you think is responsible for this tradition, whose idea was this tradition?*

M: Our great-great-grandmothers.

AW: *And why do you think they decided to do this? What is the reason for it, since there's so much pain and suffering?*

M: Our great-great-grandparents used to do it, and we don't know the reason why, or why we are still doing it and will continue to do it.

AW: *If you had the power as a woman to change this tradition, would you change it?*

M: I can't imagine that we would have the power to stop it. I don't have the power to stop it, but if I did, I would make it stop.

AW: *Why would you like to stop it?*

M: Because of the pain.

AW: *I wanted to ask you: you know you've said that when you were a little girl this was done to you and you think that Mary, little*

*Mary, will forget the pain. I'm very curious to know if you yourself have forgotten the pain that you felt when you were a little girl.*

M:   I cannot remember the pain. I've forgotten, and little Mary will also forget.

AW:  *Do you think it made any difference in the way you felt about your mother, that she had let you go without telling you that you would be operated on by strangers?*

M:   Initially, I thought my mother had . . . had fooled me, but afterward I came to believe it was tradition.

AW:  *Did you ever discuss what had happened to you with your mother?*

M:   No.

AW:  *You know that when the sexual organs are removed women do not experience sexual pleasure the way they do when they have their organs intact. In your marriage, in your relationship with your husband, do you feel that something is missing?*

M:   I haven't experienced that. As far as I'm concerned, I'm happy with my married life and feel I'm enjoying everything.

# Pratibha Parmar and Baba Lee

*Baba Lee is an Islamic scholar who is against female genital mutilation. He is one of the very few men who speak out against this practice publicly, at conferences and on radio programs. I interviewed him in the courtyard of our hotel in Banjul.*

PP:    *Can you tell me why, as an Islamic scholar, you think that female circumcision is wrong?*

BL:    Some Islamic scholars are trying to make a link between Islam and female circumcision, which is wrong. It's a tradition that had been practiced long before Islam came to this continent. It has nothing to do with Islam. It is not mentioned in the Holy Koran; it is not mentioned in any hadith. So it cannot be stated as a *sunna* or as a *faridah*. A *sunna* is a practice based on the prophet's words and deeds, and *faridah* means an obligation.

   The majority, 93 percent, of people in Gambia are Muslims. And some scholars say you cannot be a proper Muslim woman if you are not circumcised. This is not true, but it is something which is psychologically imposed on women by some scholars. You find, for example, that most of our Mandinka people call female circumcision a *sunna*. This means the prophet asked people to do it, which is wrong. It is really an option, which means it is up to you. It has nothing to do with your standard of Islamization. Some scholars used to preach that the woman who had never been circumcised was not clean, which is not true because to be spiritually clean, from an Islamic point of view, means keeping your five daily prayers, keeping your proper ablution, keeping a clean mind, and doing good work. Cleanliness doesn't mean being operated on, as in female circumcision.

Most scholars try to justify that act by explaining that when the prophet Muhammad migrated from Mecca to Medina, he happened to find a woman, an old woman, called Um'Apiah. This woman used to carry out this operation. The prophet told the woman, "If you do this again, be very careful of the way you do it." And then he explained how to do it. Not to be extreme because of the harm it will cause.

You find most of our people who support female circumcision do so just because they belong to a tribe which keeps this practice. I also belong to such a tribe. But I feel it is not honest to try to achieve your own goals in the name of religion. It is not fair. That's why I stand very firm and work very hard teaching people to clarify this.

PP:     *Why do you think it is mostly Muslim men who say it is part of the tradition and tell women to do this?*

BL:     It is a means of suppressing women. During the occasion of the circumcision they have their own school to teach women how to be obedient, how to be subdued with men, how to carry on traditions that matter.

PP:     *You are saying it's the way in which women are trained to be under the control of men?*

BL:     Indirectly, yes. To be under the control of a man, a woman has to be very strong. You have to be very strong because you don't want to be rejected by your own society. Within societies where this is practiced, if you are not among those who undergo circumcision, you will not be accepted and will be made to feel inferior. And yet from an Islamic point of view, if a Muslim doctor—or a Christian doctor, because Islam has a high regard for Christianity, too—can prove that this practice can harm a woman or can be detrimental to the health of women, Islam will fully agree with you. Because Islam is a way of life. And being a way of life, it will never impose something detrimental on any nation, tribe, or family.

For example, every Muslim has to fast during the month of Ramadan. But Islam says if you are traveling, or if you are sick, or if you are told by a doctor—even a Christian or Jewish doctor—that if you keep the fast, that might jeopardize your life, it is fully accepted in Islam, and written in the Holy Koran, that you should not keep the fast. This is my argument. Keeping a fast is an obligation that each and every one has to practice. But if it may be detrimental to your health or your life, you are allowed to forgo the fast, and you are encouraged to give to charity instead. If there is this understanding that you should not fast if it is harmful to your health, what then of female circumcision, which was never mentioned in the Holy Koran or in any form of hadith? It is just a matter of propaganda, which people are spreading because they want circumcision taken up more and more. But it is not Islamic. It is not religious. It is not obligatory. It is not *sunna*. It's a tradition.

PP:    *How do you think, then, that this position can be challenged because it's so harmful to women and their health?*

BL:  It can be easily challenged by having honest minds. We must try to teach people, try to educate people and give them genuine proof of its harmful effects, although it will take time.

PP:  *Apart from its being something that's not in the Koran, why are you personally against it?*

BL:  One, I don't want anything to be named as Islam when it is not. Second, Islam has been mistaught. To many people, many nations, many countries, Islam has become something harmful. Some people are getting scared of Islam because they think Islam is a bloody religion. They think Islam is a religion that encourages dictatorship, that encourages forcefulness and is antidemocratic. That is not true. So I have chosen to be one of those who defends Islam, the real Islam, who educates people to understand what is Islam and what is not Islam. That's why I'm part of this movement.

PP:  *Do you think fewer people are now practicing female genital mutilation?*

BL:  Yes. In The Gambia in particular, it is becoming less widespread. Fifteen years back, you could never think of speaking against it here in this country. People would have been mad at you. Today you can go around the country, set up your table, and have a meeting with people, and they listen to you, and if you have materials to show them, they become interested. That means they have started taking on the idea.

PP: *Do you have daughters of your own? Have they been circumcised?*

BL: I am married, and I have two daughters. It's so sad what happened to my first daughter. In 1990, I traveled, I went out of the country, and my mother took my daughter and had her circumcised. When I came back, I was very upset. But tell me, what can I do with my mother? Fortunately she promised me that she will never do it again, to my other daughter. If my daughter grows up and decides to do it, it's up to her. But I don't want my daughter to feel that she had been mutilated when she was young and they did something wrong to her. And I feel the act is wrong.

Islam asks us to be very polite with our parents, to be very humble with them. Even if they are wrong, we are told not to go right out and tell them they are wrong. We have to use our own ways and means to communicate with them. So I talked to my mother, and she promised me that she would never do it again. And after all, she's my friend. My mother is my very good friend. So I was able to convince her, and I believe my second daughter will not be circumcised.

# Pratibha Parmar and a Trainee Midwife

*This woman did not want to give us her name because she was concerned about her safety. I interviewed her after the workshop with circumcisers in Banjul. She was against female genital mutilation and hoped, through her future work as a midwife, to raise women's consciousness about the harmful effects of this practice.*

Q:      *What do you think about the practice of female circumcision?*

A:      I would advise you to tackle this problem very carefully, in order to avoid embarrassment to people, because it is not impossible that this tradition will be abolished one day in our society. From generation to generation, customs tend to disappear. I myself blackened my lips, but our children have categorically refused to do that, and I am convinced that the same thing will happen with circumcision. In the past, women used to give birth at home, but now women understand that giving birth in maternity wards is more hygienic and safer.

Most probably you have noticed that some people are reluctant to discuss circumcision. It is a taboo subject, which cannot be talked of anywhere, by anyone, and that is why I would advise you to be very careful when discussing it.

Q:      *Have you been circumcised yourself, and what did you feel?*

A:      I was circumcised when I was very small, and therefore I did not feel anything. According to my mother, I was only two years old.

Q:    *Do you have children who have been circumcised?*

A:    Yes. My parents circumcised them. But I would not force anyone, especially my children, to circumcise their own children.

Q:    *Do you think that people will carry on this tradition?*

A:    I think so. Because they value their traditions, the traditions inherited from the ancestors.

Q:    *Do you see any possibilities for this tradition to be stopped?*

A:    I do not think so. If parents hear about our meeting today, they will probably insult us. They will say that we are about to disclose the deepest secrets of our society. And that is why most of the participants in the meeting today have refused to talk. And to answer your questions.

Q:    *When you become a midwife, will you tell women that this practice is harmful, especially when they give birth?*

A:    Yes. Without fail, I will tell them that circumcision makes childbirth more painful. Maybe they will understand me and abandon the practice. Some practices come from ignorance. When they have realized the harmful effects of circumcision, they will abandon it.

# Pratibha Parmar and Mama and Mam Yassin

*I interviewed these two young women in Banjul. They were articulate and angry about female genital mutilation, and have organized a committee to educate young people in their school. Their voices are the hope for the future.*

PP:   *Mama, can you tell me about the work your sister and mother are doing?*

MA:   They are part of an organization, the BAFROW and GAMCOTRAP, that is trying to fight against female circumcision, to show it is a very harmful practice—it's inhumane.

PP:   *Do you know how widespread it is at your school and how many of your own friends have been circumcised?*

MA:   Well, I'm not sure, but I think most of them are circumcised. We, the youth, have organized a committee to fight against female circumcision, to talk to young people and convince them that it should be abolished.

PP:   *What kind of responses have you had from people in your school—say, from the boys—when you discussed this?*

MA:   Yeah, the boys, most of them are negative, they say that female circumcision is in the Koran, that it should be practiced. But we're trying to convince them that it's not in the Koran. Even the prophet's children were not circumcised.

PP:     *Why do you think boys support circumcision?*

MA:     Maybe it's because they don't know about *female* circumcision. They think that it's the same as male circumcision; they don't know that it does harm to the female body.

PP:     *Mam Yassin, are you also at school? What do you think about female circumcision?*

MY:     I am in the school with Mama, doing my A levels, and I think circumcision is very wrong and it should be abolished.

PP:     *Why do you think that?*

MY:     Because first of all, it's not good to a woman's body. The part that is cut affects women very much. Mothers save lots of money for this one day when the little girls are circumcised, and they could use it in other areas of life, instead of wasting it on this day, which is so harmful.

PP:     *Are many of your friends circumcised?*

MY:     Yes, most of my friends, most of the people I know who are female, are circumcised.

PP:     *And what do they think about it?*

MY:     Parents tell them, You should be proud of it, and those that have not gone are really mocked. They laugh at you and they make fun of you. So they feel proud of having done it. And they don't know the harmful effects of it.

PP:   *Do you think that young people are changing in their attitudes about this?*

MY:   Yes; now young people are changing. People are campaigning, like Mama's mother and sister. They're all campaigning against this. And as Mama said, we are forming our groups at school, and most of them agree with us. And some of the mothers come, too.

PP:   *Do you know of any incidents that have happened to your friends?*

MA:   One friend told me that when she was six months old, she was taken to be circumcised, her mum told her. But now that she is older, when she is menstruating, her stomach gets swollen. She spends about seven days menstruating, and it's very painful.

MY:   I know one of my friends bled very seriously. But on the other hand, another friend just ran off, gave the appearance that she was not going to go through with it. But she felt she would be cast out by her friends if she didn't go, so in the end she agreed to go along with it.

PP:   *And have you had pressure on you by your family to be circumcised?*

MY:   My grandmother, yes, she still believes in female circumcision. But both my mum and my dad are against it, so they've been telling me, "Don't listen to your friends, and when you go out, don't let them tempt you into going." And my mum says, "I won't have any big ceremony for you if you do it, so either way you'd be shamed, because all your friends will be partying, and I won't be present," so I have not been tempted.

PP: *And you wouldn't want to be circumcised?*

MY: No, I wouldn't.

PP: *Why?*

MY: Because as I said before, I know the harmful effects, and I'd like my parents to use the money which would pay for the party to further my education or do something for my younger brothers and sisters, rather than waste it. And I would like to be healthy and live longer.

# Pratibha Parmar and Khady, Daniele, and Nafi

*These three women are members of the Commission for the Abolition of Sexual Mutilation in Senegal. They are activists who have been campaigning against all forms of sexual mutilation for thirteen years. We were introduced to them by Awa Thiam, who is a founding member of this commission.*

PP: *Can I ask you the name of your organization and what exactly you do?*

K: Our organization is called Commission for the Abolition of Sexual Mutilation (CAMS International). In Senegal, the movement is called Women and Society. We have several objectives: research on women's problems; how to eradicate all forms of patriarchal oppression; and research on different forms of domestic violence. And we struggle against all forms of sexual mutilation, as the name of our group indicates. Our group was created in Senegal in 1982. We have a different approach from other women's movements. Our resources do not permit us to work with working-class women. Our work is voluntary, so we organize a lot of conferences and broadcasts, and we campaign on TV, radio, and newspapers. Our approach is mainly to give information and to develop awareness of the problem. A series of conferences and discussions have been held in Dakar and in Paris, because the French section is working very well and there are exchanges between us.

In 1988, for example, we organized a conference in Paris on violent assault on young girls and women, and in Dakar in 1983 we held a conference on sexual mutilation. And in

December 1992, we organized a conference which was tre-
mendously successful, which has revolutionized a lot of
taboos in Senegal. The conference was led by Daniele, who
gave a lecture on battered women and domestic violence.

PP: *How many women are there in this organization, and how long
have you belonged to this organization yourself?*

K: We have fifteen members right now. I joined the movement in 1990. I decided to join because I had heard people
speak about the movement. I have a friend, a real friend—
like a sister: she spoke on television, she organized conferences, she had traveled and she returned, and through her
I felt involved as well with the problems of the women of
Senegal and all the women of the world. So I said I would
be active in that movement, since I did not want to be
active in politics.

PP: *And what is your definition of sexual mutilation?*

K: I think there is first of all physical mutilation and then psychological mutilation. At the psychological level, I mean
battered women, oppressed at home, women who are not
treated properly at work. And physical mutilation is the
circumcision of young girls.

PP: *You have just spoken about psychological mutilation. I believe
Daniele has just held a conference about that subject. Could you
please tell us about it?*

D: First of all, I must say that I joined the movement relatively late, when I was invited by Radine, who is a young "sister" and who knows that I am feminist, pro-feminist, without being too obsessed by a certain kind of struggle. I joined the movement in 1991—it's quite recent. Nevertheless, as I am older than the others, I have been able to develop a more global approach to the problem with them.

I think that mutilation is an important problem, which affects both women and children. It affects young girls, so I believe there are areas where we can fight for a feminist cause and at the same time help children. But the global approach seems more interesting to me. Women are denied all manner of rights. Women cannot own land, cannot have credit at the bank, cannot start a company, and cannot even travel without their husband's authorization.

The conference focused on battered women. It was extraordinary because it attracted more men than women. The audience was up to 70 percent men. The debates following the conference have brought hope, in our opinion. With the exception of a few men who tried to be disruptive, everyone recognized that there was a real problem. We were lucky enough to have a lot of newspaper coverage and to have a report in *Amina,* a women's magazine which is available throughout West Africa. After that we had an interview with the radio. We were relieved we were on the right track. Because a lot of things had been touched upon. Things which people didn't talk about but were the reality of life in Senegal. And all that allowed us to become known. Because there are very few of us in the movement. We are working women, and we cannot make ourselves available all day long for the movement.

The second drawback is our lack of support from institutions. Our association has no material means, so if we discover there is a problem in Kaolack, we have to pay our own expenses to go there. That is an enormous problem. If we can overcome this hurdle, our action will be much more widespread and we will be able to expand the association. We want to work with ordinary people again, not just people like us. We must reach women who are still living with those traditions, so that we can talk to them and convince them. I think that is one of the most important elements.

One of the objectives of our association is to press for a law banning sexual mutilation. It is something which cannot be eradicated within four days, ten days, even ten years. We must be patient and not lose courage. In Senegal in 1993, sexual mutilations are still being performed, even in circles which traditionally did not practice them previously. I have also been told that in the town of Nbour, where there is a large community belonging to the Manding group, young girls were freely offering themselves to be circumcised—even without asking their parents. One of my colleagues at work described the problem to me recently. One of his nieces has friends who have been circumcised, and she wanted to be circumcised herself. Without asking her parents' permission, she had herself circumcised. There are some myths circulating that circumcision protects women from men—that they won't get pregnant and they won't have certain problems. I found all that terribly frightening. If we want to take action, those beliefs have to first disappear. Before, this mutilation was imposed on children. Now young girls offer themselves for mutilation—that is frightening.

PP:  *What you are saying is very interesting, because we have talked to a lot of women who believe that this practice is on the wane. But what we are now saying is that the problem is acquiring a new dimension.*

D:  I think that in some places there is real progress being made. Because now, in families where it was traditionally performed—for example, among people belonging to the Tricoleur tribe, where aunts and great-aunts have been circumcised—there has been a change in this generation: the mother did not have her daughter circumcised, and the daughters won't have their daughters circumcised. I also believe that the action of men is important. I have been told of instances when the father had his daughter circumcised, even if the mother did not agree. Therefore, I believe that, on one hand, the problem is diminishing because people are aware of its danger or are opposed to it. But on the other hand, there is a kind of fashion, a social imitation. We should carry out some surveys to obtain figures and be in a position to say whether or not it is on the wane.

PP:  *What other ways have you found to raise people's awareness?*

D:  In Senegal, after this conference on battered women, there was a wave of domestic violence, and in Conac a woman decided to have a National Day of Mourning and asked to speak on the radio. Khady and myself went along and asked Senegalese women and men who believed that society should not be built on violence to dress in white, and it was a real success. Even if we had had longer to prepare, the success could not have been greater. You saw people dressed in white in schools, on buses, men and women— it was marvelous. I think that is the right way to develop awareness among people.

PP: *Khady, can I ask you a personal question? What was the attitude to sexual mutilation in your family?*

K: I am of Pula origin, slightly mixed with Bambara. My family comes from Mali, where sexual mutilation is widely practiced and parents used to have all the children circumcised. We were lucky because my father did not want it. He was an intellectual and studied a lot. He was totally opposed to the mutilation of his children. But around me, many girls were mutilated. When I was a child, I was traumatized because we were told by other people that if we wanted to be a real woman, a good woman, we had to go through with it, otherwise a husband of our own class would not accept us or we would be rejected by the in-laws. But my father remained firm, and mutilation was not common within our family.

But some of my close cousins were mutilated. We had to take part in the various ceremonies, and it was horrible because they were all in tears. It was something extremely violent and very profound. At that moment, I decided to fight against circumcision and against violence against women. We had to fight within the family, and when some of my youngest cousins became threatened with mutilation, I struggled to make everyone around me aware.

My husband is a doctor, and it was easier because we could work together and explain dangers linked to mutilation and the medical consequences, and that produced some results. Mutilations are less and less frequent in our family, though the part of the family in Mali does still practice it. I still have young cousins who are mutilated—and it is now carried out in hospitals. But it is still revolting. One of the aims of mutilation is that girls should still be virgins when they marry. But we have discovered that girls who had been mutilated got pregnant before they were married because they thought that they would be protected from pregnancy. This has made parents reflect. Pregnancy was not avoided. Virginity was not safeguarded for marriage, so there was no more use for sexual mutilation.

PP:     *So you have not been circumcised?*

K:     No, my father refused to let them circumcise me.

PP:     *Is there anyone among you who has been circumcised?*

D:     Personally, I am not circumcised. I come from the West Indies, and this practice does not exist over there. I came to Senegal to get married, and I can see the problems as an outsider. I have friends who have been circumcised, and some of them are convinced that circumcised women are the best. So I cannot speak for them. But we discuss this problem very often, and I can see how important it is.

N:    I have not been circumcised, but my younger sister, just after me, she was circumcised next door, because as in Khady's case, half of my family is from the Tricoleur tribe. So my aunts on my father's side are in favor of sexual mutilation. But as my father did not agree to it, it was carried out in a different house. My sister nearly died when she was cut. Roast nuts are applied on the wound to prevent hemorrhaging, but she had a strong reaction and her legs swelled up. My father was very upset. She underwent treatment—injections and so on. She got married before me and had children. When she gave birth, she always had problems with infections.

PP:   *Did she decide to do it of her own accord?*

N:    Yes. It was her decision; I don't know why she did it. She went to a hut next door, where it was practiced. My aunt on my father's side, who is from the Tricoleur tribe, told her to go. My mother said, Your father disagrees—but she went.

PP:   *How old was she when she decided to do it?*

N:    Not even ten.

PP:   *Did you ask her recently if she regrets it?*

N:     Yes, of course she does. Because she is the only one who has problems. She has difficult pregnancies, infections all the time.

PP:    *Does she want her daughters to be circumcised?*

N:     No, no.

# Alice Walker and Tracy Chapman

*This brief interview with Tracy Chapman was conducted at the House of Slaves on Gorée Island in Senegal. Her music is well-loved, and she herself greatly admired, all across the African continent.*

AW: *Here we are sitting on the steps of what is called the House of Slaves, and I wondered what you have been thinking about and what you've been feeling about genital mutilation?*

TC: Coming to this place probably felt like the end for so many people; there must have been so much uncertainty and fear. It just made me think about what's been done to women and that they're being mutilated, and it seems so unfathomable, and the possibility of changing it—since it's more pervasive than people think—is unfathomable, too. I was just looking out at the water and thinking: I stand here as a free person as much as it's possible in this world, and it's possible that things can change and that women will have control of their bodies and have whole bodies and have a new beginning.

# Pratibha Parmar and Alice Walker

*This was the final interview we shot for the film. It took place at the House of Slaves. I wanted Alice to talk about her impressions of traveling and meeting women in Africa. I also wanted to ask her about the experiences of enslaved women who had been mutilated.*

PP: *Alice, I'd like to start by asking you to talk a little bit about where we are at the moment and the significance of being in this house and on this island.*

AW: We're on Gorée Island, off the coast of Dakar, Senegal. This is a slave castle or slave house, a place where captured Africans were kept prior to being shipped to the United States to work on the plantations, mainly in the South.

PP: *One of the things you've often spoken of when we've talked about female genital mutilation is the absence in historical accounts of the experiences of women during slavery—when they were taken on the ships, what happened to them if they had been circumcised or genitally mutilated? Can you talk about that?*

AW: Nowhere have I seen any mention of the fact that women who were enslaved along the coast of Africa—and who came out through this particular house where we are now, for instance—were probably mutilated and infibulated. So we have to imagine what that was like for those women. Not only were they subjected to all of the cruelties that everyone else was subjected to, but in addition they had been stitched shut, so that every bodily function through the vulva had to be horrendous.

PP:   *Why do you think that this is absent from historical accounts?*

AW:   The emphasis is mostly on men who were enslaved. It's very rare for there to be lengthy work about the women and their experience. Often, when you read the accounts, it's as if only men were enslaved.

PP:   *We've now been traveling in Africa for a couple of weeks, and you've met women in The Gambia and Senegal, and we've been talking with them about female genital mutilation. What has your impression been of what you've experienced so far.*

AW:   It's remarkable to me that the suffering of the children is the thing that is least considered. Children cry in pain and terror when they feel pain and terror, and yet the elders and their parents just assume that they will forget the pain that they endure. That has very much preoccupied me. I find it quite chilling.

PP:   *Is there anything else, among impressions you've had when women have spoken to you about their experience of being mutilated, that stands out?*

AW:   Most women start out saying they forgot the pain. But when you ask, they say they would abolish it as tradition. And when you ask why, they always say because of the pain. So it's something that they have fooled themselves into thinking they have forgotten and they hope their daughters will have forgotten by the time they are their age. But of course they don't forget and the body does not forget. The body does not forget pain.

PP:    *Do you feel that in any way there is hope? Because my impression has been that there are many women in Africa who are campaigning and organizing against female genital mutilation.*

AW:    Oh, I think there is, there's a great deal of hope. There are women here who are dedicated to eradicating it, and that's very good to see.

   I had a moment while I was interviewing a circumciser that was very good for me personally. I had sat there next to her, feeling a great deal of dread and anger and hostility, and there was a moment in which we looked into each other's eyes and I could see her humanity and feel that she was a human being and that she had been tricked and indoctrinated and programmed into this line of work. At this point, I sensed the person behind the horrible activity that she is engaged in. It was so good to see that she's human that I laughed.

PP:    *And what about the ceremony we went to on the first day of filming, where the young girls who had been circumcised two weeks before were to have their coming-out celebration. What did you feel then?*

AW:    The hardest thing for me was little Mary. Little Mary is four years old, the youngest one of her group who endured the operation. I was looking at her little feet the whole time. I was very upset when they sacrificed the chicken and its blood splashed onto little Mary's feet. What will Mary do with that image? Will little Mary somehow understand, as I did, that this is surely more than just an animal sacrifice. It is a lesson. It is saying: Two weeks ago we cut out your clitoris; if you say anything against us, we will cut off your head. It's a very clear message, and it's about terrorism, torture, and control.

PP: *Do you feel that you understand why the mothers continue to do it to their daughters and why women allow it to carry on in this way?*

AW: I think a lot of it is the weight of tradition and the feeling that it's been going on for so long they can't change it. Mothers and grandmothers feel bonded to their children, and yet they perpetuate it. But I also feel that if they have someone to help them think about the pain that they themselves have suffered, are suffering, and if they love their children, they can begin to stop this. We have met people along the way who have said, "This was done to me. I would never do it to my daughter."

PP: *Do you feel in any way that coming to Africa and talking to these women and making them question why they do it is helping the process of change?*

AW: I do. I think that often when there is a very severe problem, what is needed is for the people who have to deal with it daily to realize that it is something that is understood to be a problem by people far away. That it is not secret and it's not just local, but it's something that people really are thinking about all around the world. This is a kind of psychic and spiritual support that we can give to women and children wherever they suffer.

PP:   *And do you feel even more now, after having been in Africa for a while, that the people in the West who say that we shouldn't be doing something like this because we're outsiders, both to the culture and to the communities here, are really not justified at all?*

AW:   It could have been me who was passing through this slave house three hundred years ago, mutilated and infibulated. It is my duty as a witness to myself, and to the ancestors, to take this on and to deal with it properly, to the best of my ability. That is what concerns me. Not so much what other people feel should be said or done, but what is our responsibility. Do we have a responsibility to stop the torture of children we say we love, or not? I mean do we love African children? Or are we like the midwife who said that when she's cutting the child and the child screams she doesn't hear it? Are we expected to be deaf to those cries?

PP:   *Is there anything in particular that you would want to feel as you leave Africa?*

AW:   I would like to feel that this movement to eradicate genital mutilation will grow and that the consciousness of people will change. I hope this will happen quickly, because they don't have a lot of time. Just in terms of health, if you cause the debilitation of half your population from all the illnesses that you get from this surgery, you are undermining everything, from your economy to your education. I think that's part of what we have seen as we have traveled about this continent.

# Film Credits

## *Warrior Marks*

*Camera—Africa, U.S.A.*
NANCY SCHIESARI

*Camera—U.K.*
JEFF BAYNES
HARRIET COX
NIC KNOWLAND

*Camera Assistant—Africa*
LORRAINE LUKE

*Camera Assistants—U.S.A., U.K.*
LUCY BRISTOW
BRENDA HAGGART
DAVID MARSH
LOUISE STONER

*Steadicam Operator*
PHIL SAWYER

*Sound—Africa*
JUDY HEADMAN

*Sound—U.K., U.S.A.*
CHRISTINE FELCE
CLAUDIA KATAYANAGI

*Electrician*
STEPHANIE JOHNSON

*Grips*
MICK DUFFIELD
GLYNN FIELDING

*Assistant to Alice Walker*
DEBORAH MATTHEWS

*Production Coordinator—U.S.A.*
AARIN BIRCH

*Production Manager—Africa*
NAZILA HEDAYAT

*Production Secretary*
RACHEL WEXLER

Dancer/Choreographer
RICHELLE

Set Design
SHAHEEN HAQ

"Like the Pupil of an Eye:
Genital Mutilation and the
Sexual Blinding of Women"
WRITTEN BY ALICE WALKER

"Black Sisters, Speak Out"
BY AWA THIAM
EXTRACT READ BY DELIA
JARRETT-MACAULEY

"Something Inside So Strong"
BY LABI SIFFRE
© CHINA RECORDS LTD

"Wale Gnouma Don"
BY SALI SIDIBE
© STERN'S RECORDS

"Diaraby Nene"
BY OUMOU SANGARE
© STERN'S RECORDS

"Samburu Dance"
BY GRETCHEN LANGHELD
© GRETCHEN LANGHELD

Additional Music
PETER SPENCER

Titles
ZAB DESIGNS

Consultant
EFUA DORKENOO

Dubbing Mixer
PETER SMITH

Assistant Editor
FOLASADE OYELEYE

Editor
ANNA LIEBSCHNER

Production Manager
SOPHIE GARDINER

Executive Producers
DEBRA HAUER
ALICE WALKER

Producer and Director
PRATIBHA PARMAR

---

A Hauer Rawlence Production
in association with
Our Daughters Have Mothers, Inc.

For Channel 4

© Our Daughters Have Mothers, Inc.
1993

# Acknowledgments

‖‖‖‖‖‖‖‖‖‖‖‖‖‖‖‖‖‖‖‖‖‖‖‖‖‖‖‖‖‖‖‖‖‖‖‖‖‖‖‖‖‖‖‖‖‖‖‖‖‖‖‖‖‖‖‖‖‖‖‖‖‖‖‖‖‖‖‖‖‖‖‖‖‖‖‖‖‖‖‖‖‖‖‖‖‖‖‖‖‖‖‖‖‖‖

*Alice Walker:*

For her faith and trust in me.

For her clarity of vision, passion for honesty, and wisdom.

For our friendship, which I hope will be "as prevalent as petunias."

For giving me this opportunity to collaborate on something so valuable and so important.

May this be the beginning of many more creative collaborations that bring us one step closer to our dreams and our visions.


*June Jordan:*

For her rage, courage, and tenderness, which will continue to inspire me always.

For enriching my life with her abiding friendship.

For guiding me with her insightful and valuable comments about my journal entries and fragments of thought.


*Shaheen Haq:*

For being my soul mate and collaborator in life for the last twelve years.

For believing in me when I didn't.

For her unwavering love and tenderness, which I hope I never
    have to live without.
Without her, my life would not be so rich and fulfilling.


*A million thanks also to:*
Wendy Coburn, for her invaluable support and feedback during
the final stages of writing my journals, for taking me fishing to
the west coast of Canada, and for her sweet loving.

Jane Dibblin, for copyediting and verifying the translations of all
the interviews in this book before submission to the publisher,
and for her friendship over the years.

Paul Gilroy, for always encouraging me.

Leigh Haber, for her caring, meticulous, and expert editing input to the journals. It's been a pleasure working with you.

Ruth Greenstein, for all her thorough and dedicated work on the
full manuscript. It is much appreciated.

Debra Hauer, for giving me a home for my films, and for our
special friendship.

Nazila Hedayat, for her commitment to the film and her belief
in me. I embrace you and our newfound friendship.

Derek Jarman, for being an exemplary filmmaker of integrity
and principle and for continuing to fight the odds.

Isaac Julien, for his support and friendship, and for his valuable
feedback on my films and on my journal entries.

Joan Miura, for always being so helpful, warm, and efficient.

Shaffiqu, Sanjay, and Sunita, for making me feel at home in California and for your loyalty and "deshi" love.

I thank all the crew members and the production team who have worked at various stages of the film production. Each one of you has brought a gift to this film with your special skill.

And I especially acknowledge Gulab, my mother, who together with my father fought alongside Gandhi in India by derailing the railway lines used by the occupying British soldiers. A survivor, she has lived as a migrant in three different continents in her one lifetime and fought more battles than she should have had to.

*Pratibha Parmar*

It is a pleasure to acknowledge the companion spirits who've joined us in this endeavor. I must first thank my administrative assistant and goddess representative, Joan Miura, for all her work. Words cannot express my indebtedness: I've lain awake nights trying to come up with some. (How about thank you and I love you?) I also thank Harcourt Brace, my publishers for over twenty-five years, for being strongly supportive of my books, this one included. I am especially happy to have the enthusiasm and good humor of my editor, Leigh Haber, and the caring expertise of Rubin Pfeffer and Vaughn Andrews. It is a delight to work with this team of thoughtful, creative souls. I also thank my agent, Wendy Weil, and my attorney, Michael Rudell, for their concerned scrutiny of all those matters, contractual or legal, that left to myself, I might overlook, not to mention never understand.

I am immensely grateful to have such savvy and upright representatives in my business life; I think there must be a walk together on some beach somewhere for the three of us in this lifetime.

I thank my mother for her courage in accepting her responsibility as a parent who made a mistake that harmed my life. This helped teach me that parents are capable of apologizing to their children, even if their children are grown. They should be given the opportunity to do so. I thank my "Beloved," for her love and patience and dedication, and for a beauty that reminds me always of the liberating power of that which is free, natural, and whole. I thank Deborah Matthews for her assistance, especially on the London shoot: it was simply great to travel with someone so at ease with herself, so in love with music, and so interested in the world. I thank Shaheen Haq for looking after us in London with the thoughtfulness of a sister: I also thank her for the stunning set she designed for our film. I thank Pratibha Parmar for responding so instantly and enthusiastically to my suggestion that we produce a film. Working together has been a joy and one that we can perhaps prolong.

Finally I must thank Bisa Niambi, Certified Massage Therapist and Reflexologist, of Atlanta, Georgia—much to her surprise, I'm sure. When I was on the road promoting *Possessing the Secret of Joy* and addressing the issue of female genital mutilation before large perplexed audiences who'd never dreamed of such a thing, I would sometimes become so exhausted I could barely move. One night in Atlanta, after a gathering at the Shrine of the Black Madonna (where I always feel so at home), Bisa Niambi arrived at my hotel, having heard of my distress from women at the local women's bookstore, and proceeded to massage me back into my body and spirit, in order that I could continue on my journey. I had signed so many books that my right hand was especially painful. After massaging me, she got down on the floor and showed me how to do finger- and hand-strengthening exercises. It was while she was demonstrating the finger stretches, pushing her hands and arms up from the floor, that I noticed the raised scars, dozens of them, on her deep-brown arms. They were old scars, and I noticed them, oddly, because they had a glowing, burnished quality. Indeed, her arms were very beautiful. I was alarmed nonetheless and could not resist asking about them. All she said was "Warrior marks." She refused to be paid for bringing me back to myself and sending me on my way. As she left the room, I was aware that someone rare had touched my life. Here, I return the energy of love to you. *Mbele Aché!* Bisa Niambi.

Alice Walker

# BISA NIAMBI, CMT, CR

CERTIFIED MASSAGE THERAPIST
CERTIFIED REFLEXOLOGIST

HOLISTIC HEALING
BY THE BALANCE OF TOUCH,
SUPPORTIVE PROCESSING
AND EMPOWERMENT.

BY APPOINTMENT (404) 642-3309

# Afterword

Shortly after she received the manuscript of *Warrior Marks,* my editor, Leigh Haber, informed me that Bisa Niambi had taken her own life and the life of her former lover, Venus Landin, also a beautiful, caring, African-American woman. She was survived by a six-year-old daughter. I subsequently learned that Bisa had been "troubled" for many years, that there had been allegations that she'd suffered severe childhood abuse at the hands of an older brother. The scars on her arms, which I had thought were possibly made by a whip, were, apparently, self-inflicted.

## *To a Fallen Warrior*

Bisa Niambi
It could not be
clearer
that you named
renamed
yourself
and that your body too
you sought to
rearrange.

Lacking
the resources
of child-abused
popular stars
you took the razor
to yourself
believing you must
separate
the body you had
from the memory
of its suffering
at all costs.
I, who walked
half dead
so many stone-
gray days
my pockets filled
with razors
understand this.

What happens
when we come alive
to our dishonor?
What happens
when we come alive
to our rage?
Who is there to
acknowledge
the wound?
Who is there to
cradle
the hurt?

Bisa Niambi
Sister of the healing
hands
that shot your lover
and yourself
*Warrior Marks*
was to be
a gift to you.
An embrace
across time and space.
I even imagined
you opening
the book
and smiling
or shrieking
as you saw
your name
and understood
I could not
forget you.

That I knew
we were part
of a circle
of loving energy
that would sustain us
as long as we
sustained it.

Bisa Niambi
How can I now think
of your falling?

Now when women
all over the world
are coming awake?
How am I to tell
your sisters
that you
went down?
And that we might not
even have known
that you fell?
Like all the
silenced ones
of Africa
and
America
over these six
thousand
or
five hundred
years.

Bisa Niambi
Who were you?
To what Africa
through your "African" name
were you
hoping to return?
And how would it
receive you?
I have now
truly seen
the Motherland

and She lies bleeding
from coast
to coast.

O Sister,
<u>why did I not</u>
<u>massage you</u>
and teach you
how to strengthen
your soul
as you
taught me
to strengthen
my fingers
and my arms?

Bisa Niambi
These tears
from an exhausted
ocean
and a dry well
are my pilgrimage
my small circle
of stones.
For I promise you
we, your sisters,
will go on.
But we will
not go on
forgetting you.
We will go on
in this world

of treachery
and danger
supporting the standing
affirming the fallen.
And we will learn the
true meaning of your scars
and your name.
And we will make sure
your daughter knows it too.

You were a warrior
and you knew it;
no apology and no
explanation.
It is a lesson
I will never
forget.
And though
you were driven
to leave us
in a scream
of violence,
that is only
the other side
—we understand—
beloved sister
of your peaceful
tenderness.

*Mbele Aché (Forward energy—i.e., Go ahead on!)*
*Sister Bisa Niambi*

# Female Genitalia

The vestibular bulbs and circumvaginal plexus (a network of nerves, veins, and arteries) constitute the major erectile bodies in women. These underlying structures are homologous to, and about the same size as, the penis of a man. They become engorged (swollen) in the same way that a penis does.

When fully engorged, the clitoral system as a whole is roughly 30 times as large as the external clitoral glans and shaft—what we commonly know as the "clitoris."

Women's sex organs, though internal and not as easily visible as men's, expand during arousal to approximately the same volume as an erect penis.

. . . In short, the only real difference between men's and women's erections is that men's are on the outside of their bodies, while women's are on the inside.

Clitoral stimulation evokes female orgasm, which takes place deeper in the body, around the vagina and other structures, just as stimulation of the tip of the male penis evokes male orgasm, which takes place inside the lower body of the male.

—Shere Hite
*The Hite Report*

**Normal Adolescent Vulva**
**in extension**

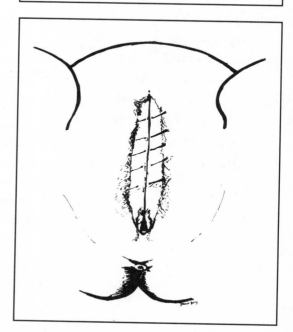

Mons

Prepuce or
hood of clitoris

Clitoris

Labia Majora

Urinary opening
or Meatus

Labia Minora

Vaginal opening

Perineum

Anus

**Infibulated Vulva**

*(Used by permission of the Minority Rights Group)*

# Types of Female Genital Mutilation

*Circumcision or Sunna:* Removal of the prepuce or hood of the clitoris, with the body of the clitoris remaining intact.

*Excision or Clitoridectomy:* Removal of the clitoris and all or part of the labia minora.

*Intermediate:* Removal of the clitoris, all or part of the labia minora, and sometimes part of the labia majora.

*Infibulation or Pharaonic:* Removal of the clitoris, the labia minora, and much of the labia majora. The remaining sides of the vulva are stitched together to close up the vagina, except for a small opening, which is preserved with slivers of wood or matchsticks.

# Contact Organizations and Advocacy Groups

*Anti-Slavery International*
180 Brixton Road
London SW9 GAT, U.K.

*Association of Nigerian Nurses and*
*Nurse Midwifes*
66 Oduduwa Way
Gra Ikeja, P.O. Box 3857
Ikeja, Lagos, Nigeria

*Babikir Badri Association for Research*
*on Women*
Ahfad University for Women
P.O. Box 167
Omdurman, Sudan

*BAFROW (Foundation for Research on*
*Women's Health, Productivity, and*
*Development)*
147 Tafsir Demba M'Bye Road
Tobacco Road Estate
P.O. Box S.K. 2854
Kanifing, The Gambia

*Cairo Family Planning Association*
50 Gomhoria Street
Cairo, Egypt

*CAMS (Commission Internationale pour*
*l'Abolition des Mutilations Sexuelles)*
B.P. 811
Dakar, Senegal

*CAMS—France*
6 Place Saint Germaine des Prés
75006 Paris, France

*Canadian Centre for Victims of Torture*
40 Westmoreland Avenue
Toronto, Ontario MGH 227, Canada

*FORWARD International*
*(Foundation for Women's Health,*
*Research and Development)*
Africa Centre
38 King Street
London WC2E 8JT, U.K.

FORWARD—*Nederland*
Beethovenstraat 123$^{IV}$
1077 JA Amsterdam, the Netherlands

GAMS *(Groupe Femme pour*
*l'Abolition des Mutilations Sexuelles)*
8 Cite Prost
75011 Paris, France

*Horn of Africa Resource and*
*Research Centre*
*c/o Family Services Centre—Ottawa-*
*Carleton*
119 Ross Avenue
Ottawa, Ontario K1Y 0NG, Canada

IAC *(Inter-African Committee on*
*Traditional Practices Affecting the Health*
*of Women and Children)*
147 Rue de Lausanne
CH-1202, Geneva, Switzerland

*London Black Women's Health*
*Action Project*
1 Cornwall Avenue
London E2 0HW, U.K.

*Minority Rights Group International*
379 Brixton Road
London SW9 7DE, U.K.

MYWO *(Maendeleo Ya Wanawake*
*Organization)*
Maendeleo House, P.O. Box 44412
Monrovia Street, 4th Floor
Nairobi, Kenya

PAI *(Population Action International)*
1120 19th Street, NW
Suite 550
Washington, D.C. 20036

*Rädda Barnen*
S-107 88 Stockholm
Torsgatan 4, Sweden

*Socialstyrelsen, Individuals and Family*
S-106 30 Stockholm, Sweden

UN *Working Group on Traditional*
*Practices*
United Nations Human Rights
Center
Palais des Nations
CH-1202, Geneva, Switzerland

*Women's Health in Women's Hands*
344 Dupont Street
Suite 106
Toronto, Ontario M5R 1V9, Canada

WIN *News (Women's International Network)*
187 Grant Street
Lexington, MA 02173

# Selected Bibliography and Suggested Reading

Armah, Ayi Kwei. *The Beautiful Ones Are Not Yet Born*. Portsmouth, N.H.: Heinemann Ed., 1989.

Boston Women's Health Book Collective. *The New Our Bodies, Ourselves*. New York: Touchstone, 1992.

Cesaire, Aime. *Discourse on Colonialism*. Translated by Joan Pinkham. New York/London: Monthly Review Press, 1972.

David, Alan. *Infibulation en République de Djibouti*. Thesis No. 131, Université de Bordeaux, publ. en 1978 par l'Amicale des Étudiants en Médecine de Bordeaux.

Dorkenoo, Efua. *Tradition! Tradition!* London: FORWARD Ltd, 1992.

————, and Scilla Elworthy. *Female Genital Mutilation: Proposals for Change*. London: Minority Rights Group International, 1992.

El Dareer, Asma. *Woman, Why Do You Weep?* London: Zed Press, 1982.

FORWARD Ltd. *Female Genital Mutilation: A Counselling Guide for Professionals*. London, 1992.

————. *Report of the First Study Conference of Genital Mutilation of Girls in Europe/Western World*. London, 1993.

Griaule, Marcel. *Dieu d'eau: Conversation with Ogotemmêli: An Introduction to Dogon Religious Ideas*. London: International African Institute, 1965.

Hedley, Rodney, and Efua Dorkenoo. *Child Protection and Female Genital Mutilation*. London: FORWARD Ltd, 1992.

Hite, Shere. *The Hite Report*. New York: Macmillan, 1976.

Hosken, Fran. *The Hosken Report: Genital and Sexual Mutilation of Females*. Lexington, Mass.: Women's International Network News, 1983.

Jordan, June. *Technical Difficulties: African American Notes on the State of the Union*. New York: Pantheon, 1991.

Kaplan, Helen. *The New Sex Therapy*. New York: Brunner/Mazel, 1974.

Kenyatta, Jomo. *Facing Mount Kenya: The Tribal Life of the Kikuyu*. New York: Random House, 1975.

Koso-Thomas, Olayinka. *The Circumcision of Women: A Strategy for Eradication*. London: Zed Press, 1992.

Lanker, Brian. *I Dream a World*. New York: Stewart, Tabori & Chang, 1989.

Lantier, Jacques. *La Cité Magique et Magie en Afrique Noire*. Paris: Editions Fayard, 1972.

Lightfoot-Klein, Hanny. *Prisoners of Ritual: An Odyssey into Female Genital Mutilation in Africa*. New York: The Haworth Press, 1989.

Masters, William, and Virginia Johnson. *Human Sexual Response*. Boston: Little, Brown, 1966.

Sherfey, Mary Jane. *The Nature and Evolution of Female Sexuality*. New York: Vintage Books, 1973.

Thiam, Awa. *Black Sisters, Speak Out: Feminism and Oppression in Black Africa*. London: Pluto Press, 1986.

Toubia, Nahid. *Female Genital Mutilation: A Call for Global Action*. New York: Women, Ink., 1993.

Walker, Alice. *Possessing the Secret of Joy*. New York: Harcourt Brace Jovanovich, 1992.

# Permissions and Photo Credits

IIIIIIIIIIIIIIIIIIIIIIIIIIIIIIIIIIIIIIIIIIIIIIIIIIIIIIIIIIIIIIIIIIIIIIIIIIIIIIIIIIIIIIIIIIIIIIIIIIIIIIIIIIIIIIIIIIIIIIIIIIIIIIIIIIIIIIIIIIIIIIIIIIIIIIIIIIII

## Permissions

The epigraph to Alice Walker's preface is from *Lovingkindness: The Revolutionary Art of Happiness*. Copyright © 1995 by Sharon Salzberg. Reprinted by permission of Shambhala Publications, Inc.

The excerpts on pages 21–24 and 209–10 are from *Possessing the Secret of Joy*. Copyright © 1992 by Alice Walker. Reprinted by permission of Harcourt Brace & Company.

The screenplay on pages 15–19, the interviews in part three, and the photographs by Deborah Matthews copyright © 1993 by Our Daughters Have Mothers, Inc. Used by permission.

The excerpts on pages 104–6 are from *Black Sisters, Speak Out: Feminism and Oppression in Black Africa* by Awa Thiam. Copyright © 1978 by Editions Denoel. Translation copyright © 1986 by Dorothy S. Blair. Used by permission of Pluto Press, London.

The quotation on page 136 is from *She Had Some Horses* by Joy Harjo. Copyright © 1983 by Joy Harjo. Used by permission of the publisher, Thunder's Mouth Press.

The lyrics on pages 186–87 are from "(Something Inside) So Strong," written and composed by Labi Siffre. Copyright © 1986 by Empire Music Ltd./Xavier Music Ltd. Used by permission.

The quotations on page 365 are from *The Hite Report* by Shere Hite. Copyright © 1976 by Shere Hite. Reprinted with the permission of Macmillan Publishing Company.

## Photo Credits

Shaheen Haq: frontispiece, pages 226 (top and bottom), 227; Ernest Harsch/Pathfinder: page 78 (top); Nazila Hedayat: pages 217 (top), 291; Deborah Matthews: pages v, 27, 31, 40, 45, 47, 51, 75 (all), 76 (bottom), 86, 120 (right and left), 122 (top and bottom), 146 (top and bottom), 172, 174, 175, 180 (top and bottom), 183 (top and bottom), 188, 210, 218, 241, 255, 301, 310, 320; Pratibha Parmar: pages 68, 76 (top), 283; Kerin Sandhu, page 132 (top and bottom); Nancy Schiesari: pages 78 (bottom), 84, 217 (bottom); Chris van Houts, page 36.